Test Yourself

Genetics

Joseph Slowinski, Ph.D.
Department of Herpetology
California Academy of Sciences
San Francisco, CA

Contributing Editors

Patrick K. Pfaffle, Ph.D.
Department of Biology
Carthage College
Kenosha, WI

Larry Hilburn, Ph.D.
Department of Biological Sciences
Arkansas State University
Jonesboro, AR

Michael Kron, M.D.
Division of Infectious Diseases
Michigan State University
East Lansing, MI

NTC LearningWorks
NTC/Contemporary Publishing Group

Library of Congress Cataloging-in-Publication Data

Slowinski, Joseph.
     Genetics / Joseph Slowinski ; contributing editors, Patrick K.
Pfaffle, Larry Hilburn, Michael Kron.
         p.   cm. — (Test yourself)
     ISBN 0-8442-2387-5
     1. Genetics—Examinations, questions, etc.   I. Pfaffle, Patrick
K. II. Hilburn, Larry. III. Kron, Michael. IV. Title. V. Series:
Test yourself (Lincolnwood, Ill.)
     QH440.3.S58   1998
     576.5'076—dc21   98-3918
                                        CIP

A *Test Yourself Books, Inc.* Project

Published by NTC LearningWorks
A division of NTC/Contemporary Publishing Group, Inc.
4255 West Touhy Avenue, Lincolnwood (Chicago), Illinois 60646-1975 U.S.A.
Copyright © 1998 by NTC/Contemporary Publishing Group, Inc.
All rights reserved. No part of this book may be reproduced, stored
in a retrieval system, or transmitted in any form or by any means,
electronic, mechanical, photocopying, recording, or otherwise, without
the prior permission of NTC/Contemporary Publishing Group, Inc.
Printed in the United States of America
International Standard Book Number: 0-8442-2387-5
18 17 16 15 14 13 12 11 10 9 8 7 6 5 4 3 2 1

Contents

Preface

The test questions and answers in this book have been written to help you more fully understand the material in an introductory genetics course. Working through these test questions prior to a test will help you pinpoint areas that require further study. Each answer is followed by a parenthetical topic description that corresponds to a section in your genetics textbook. Return to your textbook as necessary to reread and review areas in which you feel unsure about your work. See your professor for additional help, and/or get together with a study group to complete your preparation for each test. Work some problems from this test book every day. This will allow you time to prepare properly for your exams.

Good luck with your study of genetics!

Joseph Slowinski, Ph.D.

How to Use This Book

This "Test Yourself" book is part of a unique series designed to help you improve your test scores on almost any type of examination you will face. Too often, you will study for a test—quiz, midterm, or final—and come away with a score that is lower than anticipated. Why? Because there is no way for you to really know how well you understand a topic until you've taken a test. The *purpose* of the test, after all, is to test your complete understanding of the material.

The "Test Yourself" series offers you a way to improve your scores and to actually test your knowledge at the time you use this book. Consider each chapter a diagnostic pretest in a specific topic. Answer the questions, check your answers, and then give yourself a grade. Then, and only then, will you know where your strengths and, more importantly, weaknesses are. Once these areas are identified, you can strategically focus your study on those topics that need additional work.

Each book in this series presents a specific subject in an organized manner, and although each "Test Yourself" chapter may not correspond to exactly the same chapter in your textbook, you should have little difficulty in locating the specific topic you are studying. Written by educators in the field, each book is designed to correspond as much as possible to the leading textbooks. This means that you can feel confident in using this book that regardless of your textbook, professor, or school, you will be much better prepared for anything you will encounter on your test.

Each chapter has four parts:

 Brief Yourself. All chapters contain a brief overview of the topic that is intended to give you a more thorough understanding of the material with which you need to be familiar. Sometimes this information is presented at the beginning of the chapter, and sometimes it flows throughout the chapter, to review your understanding of various *units* within the chapter.

 Test Yourself. Each chapter covers a specific topic corresponding to one that you will find in your textbook. Answer the questions, either on a separate page or directly in the book, if there is room.

 Check Yourself. Check your answers. Every question is fully answered and explained. These answers will be the key to your increased understanding. If you answered the question incorrectly, read the explanations to *learn* and *understand* the material. You will note that at the end of every answer you will be referred to a specific subtopic within that chapter, so you can focus your studying and prepare more efficiently.

 Grade Yourself. At the end of each chapter is a self-diagnostic key. By indicating on this form the numbers of those questions you answered incorrectly, you will have a clear picture of your weak areas.

There are no secrets to test success. Only good preparation can guarantee higher grades. By utilizing this "Test Yourself" book, you will have a better chance of improving your scores and understanding the subject more fully.

Introduction to Genetics

1

 Brief Yourself

Genetics

Genetics is the study of genes, which are the molecular units that form the blueprint for the development of all organisms. Genes are comprised of deoxyribonucleic acid or DNA, a polymer of molecular subunits called nucleotides. The locus for a gene refers to the location on the chromosome of a particular gene. In DNA, there are four nucleotides, abbreviated A, T, C, and G. A single molecule of DNA inside a cell is called a chromosome. The sum total of genes for an organism is called the genome. Genes (with a few exceptions) encode the instructions for making proteins, which are polymers of amino acids. The process of making a protein from the instructions in genes is called translation, which involves another type of nucleic acid called ribonucleic acid or RNA. Proteins accomplish an enormous number of tasks in cells, but most are enzymes, which accelerate chemical reactions. The study of genetics falls along three subdisciplines:

– classical genetics is the study of the transmission of genes from parent to offspring = heredity;

– molecular genetics is the study of genes and their function at the molecular level; and

– population genetics is the study of the dynamics of gene frequencies in populations.

Chromosomes

Chromosomes are single DNA molecules associated with several types of protein. The combination of DNA and proteins is called chromatin. Ploidy refers to the number of sets of chromosomes possessed by a species. Most eucaryotes are diploid, possessing two sets of chromosomes, each set coming from a separate parent. A pair of chromosomes carrying the same genes, but coming from different parents, are termed homologous.

Procaryotes and Eucaryotes

There are two primary types of cell organization. Procaryotic cells are possessed by the kingdom Monera (= bacteria), and are relatively simple cells. Eucaryotic cells are more complex than procaryotic cells, from which they arose, and characterize all four remaining kingdoms of life (Protista, Fungi, Plantae, Animalia). The chief difference between procaryotic and eucaryotic cells involves the internal organization of the cell. Eucaryotic cells are compartmentalized by a variety of membrane-bound organelles, the most important of which is the nucleus. Additional eucaryotic organelles include the mitochondrion, chloroplasts (photosynthetic eucaryotes only), the endoplasmic reticulum, and the Golgi apparatus. Other important differences between procaryotes and eucaryotes relate to the chromosomes. Procaryotic cells have a single, circular, small chromosome. Eucaryotes generally have several chromosomes, each of which is linear and larger than a procaryotic chromosome.

Test Yourself

1. Who founded the modern field of genetics?

2. Who coined the word gene?

3. How many genes are contained in a genome?

4. What is a gene?

5. How do genes operate?

6. Do all genes code for proteins?

7. What are the functions of proteins?

8. Is all the DNA inside a cell part of a gene?

9. Do genes overlap?

10. Is a particular gene exactly the same in all individuals in a species?

11. What do the terms genotype and phenotype mean?

12. Do the genes of an organism entirely determine the phenotype?

13. Why are no two individuals exactly alike in any species?

14. Will two identical twins raised in the same environment be phenotypically identical?

15. How many chromosomes are there in a eucaryotic cell?

16. How many chromosomes are there in a gamete?

17. Are all species diploid?

18. What is DNA sequencing?

19. What is the Human Genome Project?

20. How is genetics relevant to medicine?

21. What are some examples of genetic disease?

22. What are the major categories of genetic disease?

23. What causes genetic disease?

24. What is a mutation?

25. Are all mutations harmful?

26. How does Down's syndrome result?

27. What is sickle-cell anemia?

28. How does sickle-cell anemia arise?

29. Under what condition will a carrier of the sickle-cell allele have the disease?

30. What can genetics do to help those with genetic disease?

31. What is genetic counseling?

32. What is gene therapy?

33. How is genetics relevant to agriculture?

34. What is artificial selection?

35. What is genetic engineering?

36. How is genetics relevant to conservation biology?

37. What are viruses?

38. Why are bacteria important to genetics?

39. How did eucaryotes arise?

40. Is the nuclear genome the only DNA in eucaryotic cells?

41. Do extranuclear genomes function the same as the nuclear genome?

42. How does procaryotic DNA differ from eucaryotic DNA?

 # Check Yourself

1. European monk Gregor Mendel (1822–1884). (**Founders of genetics**)

2. Danish plant breeder W. Johannsen in 1909. (**Founders of genetics**)

3. The number of genes varies enormously depending on the species. In humans, it is estimated that 100,000 genes comprise the genome. (**Basic molecular genetics**)

4. A gene is a segment of contiguous nucleotides along a DNA molecule that usually codes for a protein. (**Basic molecular genetics**)

5. Genes operate indirectly; they direct the formation of proteins, which then carry out various tasks in cells. (**Basic molecular genetics**)

6. No. Some genes are RNA genes, which code for RNA molecules, used in the process of translation. (**Basic molecular genetics**)

7. Proteins have an enormous number of functions. Some of the more important functions are the following: enzymes, gas transport, movement, immunity, cell receptors, hormones, cytoskeleton. (**Basic molecular genetics**)

8. No. In eucaryotes, most of the DNA in chromosomes is non-genic. In procaryotes, there is much less non-genic DNA. (**Basic molecular genetics**)

9. Almost never. There are some rare cases of overlapping genes (genes that share nucleotides). (**Basic molecular genetics**)

10. No. Genes can have variable forms called alleles. A diploid individual with identical alleles is homozygous; if the alleles are different, the individual is heterozygous. Genetic variation is one of the most important themes in biology. (**Basic Mendelian concepts**)

11. The genotype is the set of alleles possessed by an individual. The phenotype is the outward physical expression resulting from the genotype and its interaction with the environment. (**Basic Mendelian concepts**)

12. No. The environment is always a factor to some extent in the development of the phenotype. (**Formation of the phenotype**)

13. This is the result of two factors: first, considering all genes, the chance that two individuals will share an identical genotype is infinitesimal. Second, no two individuals will ever be raised in exactly the same environment. (**Formation of the phenotype**)

14. No, because there is a third factor in the development of the phenotype in addition to the genotype and environment. This third factor is called developmental noise, and represents small variations in development that arise purely from chance. However, variations produced by developmental noise will be very similar. (**Formation of the phenotype**)

15. This depends on the particular species. In humans, there are 46 chromosomes, representing 23 unique chromosomes, each present twice. (**Chromosomes**)

16. In humans, the diploid number of chromosomes is 46. But the gametes are haploid, and therefore carry 23 chromosomes. It is critical to realize that the gametes will always carry half the number of chromosomes that are found in a somatic cell (a non-sex cell). (**Chromosomes**)

17. No. Ploidy refers to the number of sets of chromosomes a species possesses in each cell. Monerans and many eucaryotes are haploid (or monoploid). Many other eucaryotes, including most animals, are diploid. Some eucaryotes are triploid, tetraploid, etc. (**Chromosomes**)

18. DNA sequencing is an enormously important method that identifies the exact sequence of nucleotides for a gene or regions between genes. (**Basic molecular genetics**)

19. The Human Genome Poject is an effort by many scientists to map and sequence the entire human genome. The project is being administered by the National Institutues of Health (NIH). (**Basic molecular genetics**)

20. Very relevant. Genetic disease such as muscular dystrophy is a very important source of disease. (**Genetic disease**)

21. Muscular dystrophy, sickle-cell anemia, hemophilia, albinism. (**Genetic disease**)

22. Genetic disease falls into three categories:

    a. inherited genetic disease, diseases like muscular dystrophy that are due to inheritance of an allele that codes for a defective protein,

    b. somatic genetic disease, diseases such as cancer that are due to mutations to genes in somatic cells that arise during one's lifetime, and

    c. chromosomal aberrations, diseases due to abnormalities in chromosome structure or number. (**Genetic disease**)

23. Genetic disease (with the exception of aberrations in chromosome number) results from mutations. **(Genetic disease)**

24. A mutation is any heritable change to DNA. Mutations can result from purely chance events or from environmental agents. **(Genetic disease)**

25. No. Most mutations are harmful because they impair the function of the proteins that genes code for, but once in a while a beneficial mutation arises. These are the changes upon which evolution by natural selection operates. **(Genetic disease)**

26. Down's syndrome is a genetic disease due to an aberration in chromosome number. Specifically, it is due to three 21st chromosomes instead of the normal two copies. **(Genetic disease)**

27. Sickle-cell anemia is a genetic disease involving a defective form of hemoglobin called hemoglobin S. When hemoglobin S is exposed to low levels of oxygen, the molecule tends to precipitate within red blood cells causing the cells to assume a sickle-like shape. This leads to many symptoms, including anemia, organ damage, and physical weakness. The disease is most common in people of African descent. **(Genetic disease)**

28. Sickle-cell anemia results from a mutation to one of the genes that code for the globin proteins, which form hemoglobin. The mutation substitutes one amino acid for another, underscoring the potential disastrous effects that can result from small alterations to proteins. **(Genetic disease)**

29. The allele that codes for the defective globin is recessive, meaning that if a normal allele is present, it will mask the defective allele. Therefore, one must be homozygous for the recessive allele to have sickle-cell anemia. This is generally true of genetic disease. **(Genetic disease)**

30. Three things: First, through therapy involving administration of the normal protein produced by recombinant DNA technology. For example, people with insulin-dependent diabetes rely upon regular injections of insulin, much of which is produced in genetically engineered bacteria. Second, through genetic counseling of couples planning families. And third, by gene therapy. **(Genetic disease)**

31. Genetic counseling is a process whereby an expert examines familial histories (= pedigrees) to determine the probability that men and women are carriers of disease-causing alleles, and the probability that their offspring will possess genetic disease. **(Genetic disease)**

32. Gene therapy is a method still in its infancy that is designed to cure genetic disease by actually replacing disease-causing alleles with normal alleles. This method holds great promise, but much work remains to be done. **(Genetic disease)**

33. In two ways: through artificial selection, and through genetic engineering. **(Non-medical applications of genetics)**

34. Artificial selection is the process of producing evolution in domesticated species under the direction of man. Artificial selection relies upon the field of quantitative genetics, which makes detailed predictions on how phenotypic traits can be expected to respond to artificial selection. **(Non-medical applications of genetics)**

35. Genetic engineering involves the production of transgenic organisms (organisms possessing foreign genes). Thus, genes for pathogen resistance, for example, can be inserted into economically important crop plants. **(Non-medical applications of genetics)**

36. Through the subdiscipline of population genetics. The conservation of rare species requires the maintenance of genetic diversity (allelic variation). Population genetics makes specific predictions on how levels of genetic diversity will respond to various environmental factors. (**Non-medical applications of genetics**)

37. Viruses are very unusual organisms that are not classified within the usual five-kingdom system because they are non-cellular. They are all cell parasites that make their living by injecting their genetic material (DNA or RNA) into living cells. The virus genetic material then directs the host cell to form new viruses, which results in the death of the host cell. (**Viruses**)

38. Bacteria are important in genetics for many reasons. Many of the principles of classical genetics are based on bacteria. More recently, most of what we understand about molecular genetics is based on bacteria. The reason for this is that bacteria are simpler than eucaryotes and hence easier to study. (**Bacteria**)

39. Eucaryotes arose from endosymbiosis involving multiple procaryotic species. Endosymbiosis is a symbiosis involving the living within a cell of another, smaller cell. Mitochondria and chloroplasts represent the vestiges of small procaryotes that took up residence in larger procaryotic cells. Recent evidence suggests that the nucleus also may represent an ancient endosymbiotic event. (**Eucaryotes**)

40. No. Both the mitochondrion and chloroplast possess chromosomes, which are circular as one might expect given the endosymbiotic origin of those organelles from bacteria. A genome existing outside the nucleus in eucaryotes is called an extranuclear genome. (**Extranuclear genomes**)

41. Almost. Extranuclear genomes direct the formation of proteins as does the nuclear genome, but there are some differences. For one, extranuclear genomes use their own ribosomes (the particles that cells use to form proteins). Further, extranuclear genomes are more similar in organization to procaryotic DNA than to nuclear DNA. (**Extranuclear genomes**)

42. First, eucaryotic DNA is characterized by large amounts of non-genic DNA; procaryotic DNA on the other hand has very little wasted DNA. Also, eucaryotic DNA is associated with more protein than is procaryotic DNA. And eucaryotic genes are split genes, having both coding regions called exons, and non-coding regions called introns. Procaryotic genes are almost never split. (**Procaryotes vs eucaryotes**)

Grade Yourself

Circle the numbers of the questions you missed, then fill in the total incorrect for each topic. If you answered more than three questions incorrectly, you need to focus on that topic. (If a topic has less than three questions and you had at least one wrong, we suggest you study that topic also. Read your textbook, a review book, or ask your teacher for help.)

Subject: Introduction to Genetics

Topic	Question Numbers	Number Incorrect
Founders of genetics	1, 2	
Basic molecular genetics	3, 4, 5, 6, 7, 8, 9, 18, 19	
Basic Mendelian concepts	10, 11	
Formation of the phenotype	12, 13, 14	
Chromosomes	15, 16, 17	
Genetic disease	20, 21, 22, 23, 24, 25, 26, 27, 28, 29, 30, 31, 32	
Non-medical applications of genetics	33, 34, 35, 36	
Viruses	37	
Bacteria	38	
Eucaryotes	39	
Extranuclear genomes	40, 41	
Procaryotes vs eucaryotes	42	

Cell Division and Life Cycles

2

Brief Yourself

Eucaryotic Cell Cycle

The eucaryotic cell cycle consists of four consecutive phases: G_1, S, G_2, and M. G_1 comes immediately after cell division and is characterized by a period of rapid cell growth. During the S phase, the chromosomes are replicated giving rise to sister chromatids, a necessary preparation for cell division. G_2 is characterized by biochemical changes in preparation for division. G_1, S, G_2 are collectively referred to as interphase. Finally, M or mitotic cell division begins. Cell division is divided into two parts: mitosis, or division of the nucleus, and cytokinesis, or division of the cytoplasm. Ultimately, two new daughter cells are produced.

Overview of Mitosis and Meiosis

In eucaryotes, there are two kinds of cell reproduction, each with a different function. Mitosis is a single division event that produces identical daughter cells, whereas meiosis corresponds to two division events in rapid succession that produce haploid cells. Mitosis functions to produce new cells in development or to replace cells. It also functions as asexual reproduction in single-celled eucaryotes. Meiosis functions to produce gametes for sexual reproduction. In males, meiosis produces four haploid cells that will differentiate into four sperm cells. The entire process from meiosis to sperm is called spermatogenesis. In females, however, each division is very asymmetrical in terms of the provisioning of cytoplasm. The end result is three polar bodies and one large cell which will differentiate into the ovum. The entire process is called oogenesis.

Mitosis

Mitosis occurs in four phases. After interphase, the first phase of mitosis is prophase and is characterized by three cellular changes: the chromosomes condense, the nucleus disappears, and the spindle apparatus appears. The spindle apparatus forms from organelles called centrioles and is composed of microtubules. In metaphase, the chromosomes come to lie along the equator of the cell. In anaphase, the chromatids are separated by sliding of the spindle fibers over one another, as in muscle contraction. During telophase, the separated chromatids reach the opposite poles, and the changes of prophase are undone at this point. Usually, cytokinesis begins during anaphase, and becomes apparent in telophase.

Meiosis

Meiosis consists of two successive divisions, termed meiosis I and II. The first division is called the reduction division and it is during this division that haploidy is achieved. The second division, the equational division, is similar to mitosis. In each division, the phases bear the same names as in the analogous divisions of mitosis. A critical difference between mitosis and meioisis is that during prophase I, homologous chromosomes physically pair up and exchange blocks of DNA in the process called crossing-over. Crossing-over is a type of recombination, the joining together of DNA from different sources. Homologous chromosomes are chromosomes that carry the same genes. The state of being paired during prophase is termed synapsis, and the two connected chromosomes are collectively called a tetrad in reference to the two pairs of chromatids. During metaphase I, the paired chromosomes line up on the equator of the cell. Because of this configuration, anaphase will separate homologous chromosomes rather than sister chromatids, leading to haploidy. Between the two is an interphase without an S phase. Meiosis II involves separation of the sister-chromatids. An easy way to understand the difference between meiosis I and II is the following: meiosis I is the separation of homologous chromosomes, whereas meiosis II is the separation of sister chromatids.

Eucaryotic Life Cycles

There are three general types of life cycles among eucaryotes. Protists, fungi, and algae have a haploid life cycle, where the majority of the life cycle is spent in the haploid condition. Sexual reproduction involves fusion of gametes produced by mitosis. The zygote then undergoes meiosis to produce haploid cells starting the cycle anew. Animals have the familiar diploid cycle, where the majority of the life cycle is spent in the diploid condition. Meiosis produces haploid gametes, which fuse to form a zygote. Mitosis of the zygote produces a new multicellular individual. Plants have alternation of generations, where multicellular haploid and diploid phases are of roughly equal duration in the life cycle.

Test Yourself

1. What is the major function of mitosis?

2. If a diploid cell has 10 chromosomes before mitosis, how many will the daughter cells each possess after mitosis?

3. How does mitosis in animals differ from that in plants?

4. The two parts of mitosis are _____ followed by _____.

5. Why is it impossible for the reduction division to occur in mitosis?

6. What exactly happens when the nucleus disappears during prophase?

7. During interphase, the chromosomes are invisible under a light microscope. Why?

8. How do chromosomes condense?

9. Why do the chromosomes condense?

10. After the chromosomes are replicated during the S phase of the cell cycle, how are the chromatids held together?

11. Why does the nuclear membrane disappear during prophase?

12. What is the spindle apparatus?

13. How does the spindle apparatus form?

14. How do the spindle fibers attach to the chromosomes?

15. Do all of the spindle fibers attach to the chromosomes?

16. How is anaphase accomplished?

17. How does cytokinesis occur in animal cells?

18. How does cytokinesis occur in plants?

19. In what ways does meiosis I differ from mitosis?

20. In what ways does meiosis II differ from mitosis?

21. If a diploid cell has 10 chromosomes before meiosis, how many will the daughter cells each possess after meiosis I?

22. If a diploid cell has 10 chromosomes before meiosis, how many will the daughter cells each possess after meiosis II?

23. A chromosome may pair up with any other during prophase I. True or false?

24. Does meiosis occur to haploid cells?

25. Does meiosis occur in haploid species?

26. What might happen if a pair of homologs failed to separate during meiosis I?

27. A zygote is formed from the fusion of a sperm carrying chromosomes $A1$ and $B1$ and an egg carrying the homologs $A2$ and $B2$. What will be the configuration of chromosomes in an adult cell? $A1A1B1B1$, $A2A2B2B2$, $A1A1B2B2$, or $A1A2B1B2$?

28. What are the ultimate products of meiosis in males?

29. What are the ultimate products of meiosis in females?

30. A female is heterozygous for a gene Aa, while a male is homozygous aa. Write down the genotypes of the possible gametes each could produce.

31. Where in an animal does meiosis occur?

32. When an alga with the haploid life cycle reproduces sexually, are the offspring genetically identical to each other?

33. Would greater genetic variability be expected among the sexually produced offspring of an alga or an animal?

34. What are chiasmata?

35. Is there an upper limit to the number of times a cell can divide?

Check Yourself

1. To produce genetically identical daughter cells. (**Mitosis**)

2. 10. (**Mitosis**)

3. In two ways:

    a. most plant species lack centrioles,

    b. cytokinesis in animals involves a cleavage furrow, whereas in plants it involves formation of a cell plate. (**Mitosis**)

4. Karyokinesis, cytokinesis. (**Mitosis**)

5. Because of the way the chromosomes line up in tandem along the equator during metaphase. For a reduction division to occur, homologous chromosomes would have to be paired up so as to allow for their separation as occurs in meiosis I. (**Mitosis**)

6. The contents of the nucleus do not disappear, rather the nuclear membrane disintegrates. (**Mitosis**)

7. Because they are in an extended or relaxed state. (**Chromosomes**)

8. Chromosome condensation involves three levels of organization:

    a. first, the DNA is wrapped around octamers of histone proteins (2H2A, 2H2B, 2 H3, 2H4). H1 histones serve as linkers to connect adjacent octamers. The combination of DNA plus the octamer and H1 forms a nucleosome. Nucleosomes are probably in direct juxtaposition along the length of the DNA molecule,

    b. second, the nucleosomes are wound in a higher-order supercoil called the solenoid, and

    c. in the third level of condensation, the solenoid is somehow organized upon a protein scaffold. The details of this third level of condensation are poorly understood. (**Chromosomes**)

9. To facilitate movement of the chromosomes, which is the major theme of cell division. (**Chromosomes**)

10. Metaphase chromosomes are often represented as X's. The constriction corresponds to the centromere, which is a region of the DNA molecule where the chromatids are held together. The kinetochore is a protein body on the centromere that serves as the attachment site for the spindle fibers. (**Chromosomes**)

11. To facilitate the movement of the chromosomes to opposite poles of the cell. (**Mitosis**)

12. The spindle apparatus is a series of microtubules that attach to the kinetochores on the chromosomes and are responsible for pulling the sister-chromatids to the opposite poles of the cell. (**Mitosis**)

13. During interphase, there is a single pair of centrioles in animals, spindle organizer in plants, which divide during prophase to form an extra pair which then migrate to the other pole. The two pairs of centrioles form between them the spindle apparatus. (**Mitosis**)

14. They attach to kinetochores. (**Mitosis**)

15. No. Polar microtubules are fibers that are not attached to chromosomes. (**Mitosis**)

16. By sliding of the microtubules across one another. This separates the sister-chromatids and pulls them to the opposite poles of the cell. (**Mitosis**)

17. In animals, cytokinesis results from the formation of a cleavage furrow around the midline of the cell that is produced by bands of proteins. As these proteins contract, the cell is pinched into two new cells. (**Mitosis**)

18. Plant cells differ from animal cells in being encapsulated by a rigid cell wall. This prevents cytokinesis by furrowing. Instead, cytokinesis results from the deposition of a cell plate along the midline. (**Mitosis**)

19. Four ways:

    a. in prophase I, the homologous chromosomes enter into synapsis,

    b. crossing-over occurs during prophase I,

    c. during metaphase I, paired homologs are aligned along the equator of the cell. In metaphase of mitosis, all the chromosomes are aligned in tandem along the equator, and

    d. the homologs are separated during anaphase I, whereas anaphase of mitosis leads to separation of the sister-chromatids. Thus, meiosis I leads to haploid daughter cells. (**Meiosis**)

20. The second division of meiosis is essentially a mitotic division, but differs from mitosis in two ways:

    a. first, there is in meiosis II a haploid number of chromosomes, and

    b. second, each chromatid is potentially different from its sister-chromatid because of crossing-over. (**Meiosis**)

21. 5. (**Meiosis**)

22. 5. (**Meiosis**)

23. False. Chromosomes only synapse with homologs. (**Meiosis**)

24. No. A cell must be diploid to go through meiosis. (**Meiosis**)

25. Yes. Meiosis can and does occur in haploid species, but haploid cells must first fuse to form a zygote, which then enters meiosis. (**Meiosis**)

26. This is known as non-disjunction, and can lead to aberrations in chromosomal number, such as Down's syndrome. (**Meiosis**)

27. $A1A2B1B2$. The zygote will be $A1A2B1B2$, and every cell in the adult is produced by mitosis and is therefore genetically identical to the zygote. (**Mitosis**)

28. Sperm. (**Meiosis**)

29. Ova or egg cells. (**Meiosis**)

30. The female can produce *A* or *a* gametes. The male can only produce *a* gametes. It is critical to realize that gametes are haploid and therefore will only carry one allele from each gene. (**Meiosis**)

31. In organs called gonads. Cells that might ultimately become gametes are called germ cells; cells of the body that will never form gametes are called somatic cells. (**Meiosis**)

32. No. Crossing-over in combination with the reduction division ensures genetic variability among the haploid offspring in algal reproduction. (**Meiosis**)

33. In the animal; because the syngamy (the fusion of gametes) adds an extra dimension of variability not possessed by the alga, where the products of meiosis directly form the new organisms. (**Meiosis**)

34. Chiasmata (chiasma, s.) are the visible manifestations of crossing-over. They give the appearances of little crosses. (**Meiosis**)

35. Yes. Each time a somatic cell chromosome divides, the terminal regions of the chromosomes, called the telomeres, become a little shorter because of the inability of the cell to replicate the tips. Germ cells and cancer cells avoid this problem with an enzyme called telomerase that has the ability to catalyze addition of DNA to the telomeres. (**Mitosis**)

Grade Yourself

Circle the numbers of the questions you missed, then fill in the total incorrect for each topic. If you answered more than three questions incorrectly, you need to focus on that topic. (If a topic has less than three questions and you had at least one wrong, we suggest you study that topic also. Read your textbook, a review book, or ask your teacher for help.)

Subject: Cell Division and Life Cycles

Topic	Question Numbers	Number Incorrect
Mitosis	1, 2, 3, 4, 5, 6, 11, 12, 13, 14, 15, 16, 17, 18, 27, 35	
Chromosomes	7, 8, 9, 10	
Meiosis	19, 20, 21, 22, 23, 24, 25, 26, 28, 29, 30, 31, 32, 33, 34	

Mendel and the Basics of Genetics

3

Brief Yourself

Gregor Mendel

The modern field of genetics was founded by a monk named Gregor Mendel (1822–1884). While living in a monastery in what is today Brno of the Czech Republic, he became involved in breeding experiments with the pea plant *Pisum sativum*. Crossing parental plants that were true breeding (homozygous), he classified offspring in seven trait categories: plant height (tall vs dwarf), seed shape (round vs wrinkled), seed color (yellow vs green), flower color (purple vs white), pod shape (inflated vs constricted), pod color (yellow vs green), and flower position (axial vs terminal). From rigorous analyses of his data, he drew a number of conclusions. These conclusions, which were published in 1865, can be summarized as follows:

1. Phenotypic traits are due to factors that always maintain their own discrete identity. Prior to the twentieth century, very little was understood about the mechanisms of heredity. A commonly held but erroneous notion was that the factors of heredity simply blended together to form the phenotype, much like mixing paints of different colors together. Mendel, however, showed conclusively that heredity is not blending, but is particulate.

2. Genes come in pairs (a consequence of being diploid). Variable forms of a gene are called alleles. If an individual has two identical alleles for a gene, that individual is homozygous; heterozygous otherwise.

3. One allele tends to dominate another in the formation of the phenotype.

4. Alleles for a gene segregate randomly. Today known as The Law of Segregation.

5. Alleles at different genes assort randomly. Today known as The Law of Independent Assortment.

Because of its mathematical nature, Mendel's paper was largely ignored until the turn of the century, when it was independently rediscovered by three botanists.

Some Further Concepts

It is customary to denote alleles for a gene with single letters. Dominant alleles are usually represented with upper case letters (e.g., *A*), recessive alleles with lower case letters (e.g., *a*). The exact constitution of alleles (for one or more genes) of an individual is the genotype. The outward physical appearance resulting from the genotype and its interaction with the environment is the phenotype. Classical genetics is especially concerned with the results of crosses of individuals with known genotypes. A cross where only a single gene is the focus is a monohybrid cross; one where two genes are the focus is a dihybrid cross, etc. Pedigrees are charts that track the reproductive relationships among individuals. Geneticists use them to track the transmission of particular alleles. The following is a simple example of a pedigree involving two generations.

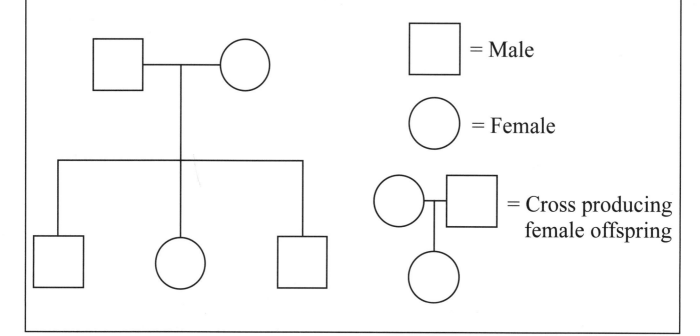

 Test Yourself

1. Give an exact definition for dominance vs. recessiveness.

2. What exactly does random segregation mean?

3. If there are two alleles for a gene *A* and *a*, how many possible genotypes will there be for a diploid species?

4. If there are two alleles for a gene *A* and *a*, how many possible genotypes will there be for a haploid species?

5. How will a haploid species differ from a diploid species in terms of dominance/recessiveness?

6. What will be the genotypic and phenotypic ratios resulting from a cross of pea plants *Tt* × *Tt*? *T* gives a tall plant; the recessive allele *t* gives a short plant.

7. What will be the genotypic and phenotypic ratios resulting from a cross of pea plants *Tt* × *tt*?

8. What will be the genotypic and phenotypic ratios for a cross of pea plants *Tt* × *TT*?

9. What do the ratios from questions 6–8 mean?

10. In a monohybrid cross involving a certain plant species, the following genotypes are observed in the offspring: 230 *AA*: 501 *Aa*: 240 *aa*. What was the original cross?

11. In a monohybrid cross involving a certain plant species, the following phenotypes are observed in the offspring: 601 "A": 630 "a". What was the original cross?

12. Explain why in many families certain genetic diseases skip generations.

13. If two haploid organisms reproduce sexually *A* × *a*, what will be the genotypic and phenotypic ratios?

14. Use the following pedigree to decide whether the trait possessed by the darkened individuals is dominant or recessive.

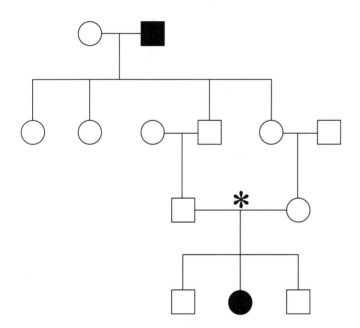

15. From the pedigree in question 14, identify as many genotypes as is possible.

16. What proportion of offspring resulting from the cross marked with an asterisk will possess the recessive trait?

17. Use the following pedigree to decide whether the trait possessed by the darkened individuals is dominant or recessive.

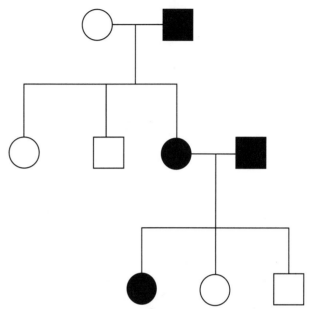

18. On the pedigree in question 17, identify as many genotypes as is possible.

19. What would be a medical use for pedigrees?

20. On the following pedigrees, the darkened individuals possess cystic fibrosis, a genetic disease due to a recessive allele. What is the probability that if individuals 1 and 2 cross, they will produce a child with cystic fibrosis?

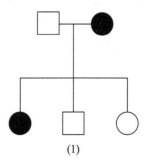

(1)

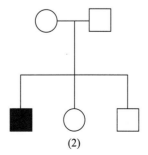

(2)

21. In a study of 33 pairs of identical twins, it was found that of 7 pairs, one child could roll the tongue whereas the other sibling could not. Explain this result.

22. Explain exactly what Mendel's Law of Independent Assortment means.

23. What is required for independent assortment to occur?

24. What feature of meiosis I results in independent assortment?

25. How many gametic genotypes are possible from an individual with the genotype *AaBb*? Assume the genes are assorting independently.

26. How many possible gametic genotypes are possible from an individual with *Aabb*? Assume the genes are assorting independently.

27. Enumerate all the possible gametes from question 25.

28. What is the probability of an *AB* gamete from question 25?

29. How many different gametes are possible from an individual *AaBbCcDd*? Assume independent assortment.

30. The classic example of a dihybrid cross is Mendel's cross between two pea plants heterozygous for genes controlling pea color and shape *YySs* × *YySs*. *Y* gives yellow peas, *y* green peas; *S* gives smooth peas, *s* wrinkled peas. Give the genotypic and phenotypic ratios resulting from this cross.

31. Do the same for *YySS* × *yySs*.

32. What is the probability of a *YySs* offspring from question 31?

33. What is the probability of a *Yyss* offspring from question 31?

34. What is the probability of a yellow, smooth offspring from question 31?

For questions 35–40, assume a cross of *Aa* × *Aa*

35. What is the probability of an *AA* followed by an *Aa*?

36. What is the probability of an *AA* **or** an *Aa*?

37. If five offspring are sampled randomly, what is the probability that 3 will be *AA* and 2 *Aa*?

38. What would be the probability of 7 *AA*, 3 *Aa*, and 2 *aa*?

39. What is the probability that of six offspring sampled randomly, at least 1 will be *aa*?

40. How many offspring would have to be sampled to have a probability of greater than 95% that at least one offspring will be *aa*?

Check Yourself

1. An allele *A* is dominant to another *a* if the homozygote *AA* shows the same phenotype as the heterozygote *Aa*. (**Basic Mendelian concepts**)

2. It means that each member of a pair of alleles has an equal probability (50%) of being represented in a gamete. (**Basic Mendelian concepts**)

3. Three: *AA*, *Aa*, and *aa*. The general formula is N = n(n + 1)/2, where N is the number of genotypes and n is the number of alleles. (**Basic Mendelian concepts**)

4. Two: *A* or *a*. It is critical to realize that for gametes or haploid organisms, only one allele from each gene will exist in the cell. (**Basic Mendelian concepts**)

5. Haploid species do not display the phenomenon of dominance/recessiveness for the simple reason that there is only one allele per gene. (**Basic Mendelian concepts**)

6. Problems of this type can be solved with a tool called a Punnett Square. The basic problem is to figure out what gametic genotypes will be produced by each parent, and then these are placed across the top and along the side of a rectangular grid:

	T	*t*
T		
t		

    The contents of the grid are then filled in:

	T	*t*
T	*TT*	*Tt*
t	*Tt*	*tt*

    The contents of the grid give the expected genotypic and phenotypic ratios from the cross, which will be 1 *TT*: 2 *Tt*: 1 *tt* or 3 tall: 1 short. Two assumptions are made by the Punnett Square: that the gametic types from each parent occur in equal frequencies (random segregation), and that the gametes fuse randomly. Neither of these assumptions will always be true. (**Monohybrid crosses**)

7. 1 *Tt*: 1 *tt* or 1 tall: 1 short. (**Monohybrid crosses**)

8. 1 *TT*: 1 *Tt* or 100% tall. (**Monohybrid crosses**)

9. They are probabilities. This means that they predict the frequency of particular genotypic and phenotypic categories if very large numbers of offspring are sampled. (**Monohybrid crosses**)

10. This is basically a 1:2:1 ratio, which results from the cross *Aa* × *Aa*. (**Monohybrid crosses**)

11. *Aa* × *aa*. (**Monohybrid crosses**)

12. Because many, probably most, genetic diseases are due to recessive alleles. For an individual to possess a recessive genetic disease, the individual must be homozygous for the allele. Thus, such diseases can skip generations as the recessive allele will often be paired with a normal allele in an offspring. **(Pedigree analysis)**

13. 1 *A*: 1 *a* or 1 "*A*": 1 "*a*". **(Monohybrid crosses)**

14. Recessive. Note how the trait skips two generations. However, one can not rely upon this rule; the only sure way to solve these problems is try both possibilities and see if one can definitely be eliminated. **(Pedigree analysis)**

15. 1st generation: *A_*(the second allele could be either *A* or *a*), *aa*; 2nd generation: *Aa, Aa, A_ Aa, Aa, A_*; 3rd generation: *Aa, Aa*; 4th generation: *A_, aa, A_*. **(Pedigree analysis)**

16. 25%. **(Pedigree analysis)**

17. Dominant. **(Pedigree analysis)**

18. 1st generation: *aa, Aa*; 2nd generation: *aa, aa, Aa, Aa*; 3rd generation: *A_, aa, aa*; **(Pedigree analysis)**

19. The most obvious medical use is to track the transmission of disease-causing alleles. This kind of information can be used in genetic counseling, where an expert can advise potential parents on the probability of producing a child with a genetic disease. **(Pedigree analysis)**

20. 25%. **(Pedigree analysis)**

21. The influence of the environment or developmental noise. **(Formation of the phenotype)**

22. It is a generalization of the Law of Segregation to two or more genes. Specifically, it means that every possible gametic genotype is equally likely. **(Independent assortment)**

23. Generally, the genes under consideration must all be on different chromosomes. However, genes far enough apart on the same chromosome can also undergo independent assortment. **(Independent assortment)**

24. Independent assortment results from the fact that no constraint exists on the orientation of tetrads on the cell's equator during metaphase I. Hence, the following two possibilities are equally likely: **(Independent assortment)**

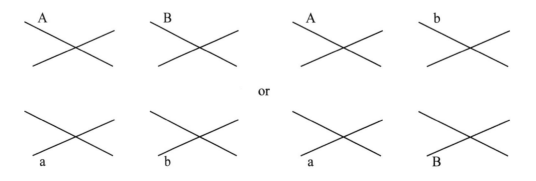

25. 4. The general formula is $N = 2^n$, where N is the number of gametic genotypes and n is the number of heterozygous genes. (**Independent assortment**)

26. 2. (**Independent assortment**)

27. *AB, Ab, aB, ab.* (**Independent assortment**)

28. 25%. (**Independent assortment**)

29. 16. (**Independent assortment**)

30. 1 *YYSS*: 2 *YYSs*: 2 *YySS*: 4 *YySs*: 1 *YYss*: 2 *Yyss*: 1*yySS*: 2 *yySs*: 1 *yyss*, or 9 yellow, smooth: 3 yellow, wrinkled: 3 green, smooth: 1 green, wrinkled. (**Dihybrid crosses**)

31. 1 *YySS*: 1 *yySS*: 1 *YySs*: 1 *yySs*, or 1 yellow, smooth: 1 green, smooth. (**Dihybrid crosses**)

32. 25%. (**Dihybrid crosses**)

33. 0 %. (**Dihybrid crosses**)

34. 50%. (**Dihybrid crosses**)

35. This uses the product rule of probability. P(*AA* and *Aa*) = P(*AA*) × P(*Aa*) = ¼ × ½ = ⅛. (**Probability**)

36. This uses the sum rule of probability. P(*AA* or *Aa*) = P(*AA*) + P(*Aa*) = ¾. (**Probability**)

37. Problems of this nature correspond to the binomial distribution and can be solved with the general equation $P(\# k = r, \# l = s) = (n!/r!s!)P(k)^r P(l)^s$, where n = r + s and n! = 1 × 2 × 3 . . . n. For this particular example, we would plug the following numbers into the equation: $(5!/3!2!)(.25)^3(.50)^2 = 0.04$. (**Probability**)

38. This problem differs from question 37 in that there are three as opposed to two outcomes. The binomial equation can easily be generalized to situations where there are more than two outcomes; the following would be the equation for a trinomial distribution: $(n!/r!s!t!)P(k)^r P(l)^s P(m)^t$. For the present problem, the desired probability is $(12!/7!3!2!)P(.25)^7 P(.50)^3 P(.25)^2 = 0.004$. (**Probability**)

39. Problems that ask "what is the probability that at least one . . . " can be solved easily by turning the problem around. The probability that of six offspring, at least one will be *aa* is the same as one minus the probability of no *aa*'s. Thus, P (at least one *aa* out of 6 offspring) = $1 - (.75)^6 = 0.82$. (**Probability**)

40. This is a fairly complicated problem. First, we note that the probability of at least one *aa* out of n offspring is $1 - (.75)^n$. We can now set up the inequality $.95 > 1 - (.75)^n$, and solve for n. This gives a value of n > 10.41, meaning that we would have to sample 11 offspring to have a probability of greater than 95% of at least one *aa*. (**Probability**)

Grade Yourself

Circle the numbers of the questions you missed, then fill in the total incorrect for each topic. If you answered more than three questions incorrectly, you need to focus on that topic. (If a topic has less than three questions and you had at least one wrong, we suggest you study that topic also. Read your textbook, a review book, or ask your teacher for help.)

Subject: Mendel and the Basics of Genetics

Topic	Question Numbers	Number Incorrect
Basic Mendelian concepts	1, 2, 3, 4, 5	
Monohybrid crosses	6, 7, 8, 9, 10, 11, 13	
Pedigree analysis	12, 14, 15, 16, 17, 18, 19, 20	
Formation of the phenotype	21	
Independent assortment	22, 23, 24, 25, 26, 27, 28, 29	
Dihybrid crosses	30, 31, 32, 33, 34	
Probability	35, 36, 37, 38, 39, 40	

Extensions to Mendelian Genetics

4

Brief Yourself

Extensions to Mendelian Genetics

In the years since the rediscovery of Mendel's seminal paper, geneticists have learned that the production of the phenotype from the genotype is often more complicated than Mendel's model. The following are extensions to Mendel's simple model:

- other forms of allelic interaction beside dominance/recessiveness

- multiple allelism

- linkage in general

- sex linkage

- one gene can affect multiple traits (pleiotropy)

- one trait can be affected by multiple genes

- variable penetrance and expressivity

- interaction among genes

- importance of the environment

Other Forms of Allelic Interaction

There are two other ways that alleles can interact in the formation of the phenotype beside dominance/recessiveness. The first is incomplete dominance, the situation where the phenotype of the heterozygote is intermediate between the phenotypes of the two homozygotes. Note that the phenotype does not have to be exactly intermediate; one allele generally tends to dominate the other to a varying extent. Dominance/recessiveness can be considered a special type of incomplete dominance, forming one end of a continuum. Semidominant allele pairs, which result in exactly intermediate phenotypes, form the other end of the continuum. Many molecular phenotypes, such as the presence of antigens in red blood cells, show codominance, the situation where the phenotype of the heterozygote is a comingling (but not a blending) of the phenotypes of the homozygotes. The ABO blood group is a classic example of codominance. The genotype $I^A I^B$ results in both A and B antigens on red blood cell membranes. Both the I^A and I^B alleles are dominant to the O allele, which does not result in an antigen.

Traits Affected by Multiple Genes

It is now well understood that most traits are affected by multiple genes, not just one. These traits range from traits such as coat color in mammals that are determined by several genes, to quantitative traits such as height in humans, that are controlled by many genes, each exerting a small effect.

Interaction Among Genes

Genes, just as alleles, can interact in the production of the phenotype. Gene interaction occurs whenever the contribution of the alleles at one locus to the phenotype depends on the alleles at other loci. Gene interaction leads to dihybrid phenotypic ratios that are non-Mendelian. Deviations from a Mendelian ratio should lead the student to suspect gene interaction.

Importance of the Environment

The importance of the environment in the formation of the phenotype can scarcely be overemphasized. Virtually all traits, whether controlled by one gene or multiple genes, will to some extent be affected by the environment.

 # Test Yourself

1. What is the difference between penetrance and expressivity?

2. Polydactyly in humans is due to a dominant mutation. Some humans with this allele do not show the trait. What is this an example of?

3. If individuals of a mammal species with a fixed genotype show a wide range of coloration, what would this be an example of?

4. Four-o'clocks are tropical American flowers. If two individuals with pink flowers are crossed, the following offspring are produced: 1 red: 2 pink: 1 white. Explain this result.

5. If white and pink four-o'clocks were crossed, predict the genotypic and phenotypic ratios of the offspring.

6. What does a non-Mendelian phenotypic ratio for a cross involving a single gene tell one?

7. What if two plants were crossed and they produced dark red, slightly less than dark red, and white flowers in the ratio 1:2:1? What is going on here?

8. In the ABO blood group individuals with the $I^A I^B$ genotype have the AB phenotype. What kind of allelic interaction does this represent?

9. Predict the genotypic and phenotypic ratios resulting from $I^A I^O \times I^A I^B$.

10. Why is AB the universal recipient?

11. Why is O the universal donor?

12. If there is a certain species of plant wherein the homozygotes for a color-controlling gene are red and white, but the heterozygote has white flowers flecked with red, what is this an example of?

13. Determine the order of dominance among the alleles at the c locus, which controls coloration in rabbits.

    c^+c = wildtype coloration

    c^+c^{ch} = wildtype coloration

    $c^{ch}c^h$ = chinchilla with black tips

    c^hc = Himalayan coloration

    cc = albinism

14. Provide an explanation for dominance/recessiveness based on your knowledge that genes code for proteins.

15. Explain albinism.

16. How do new alleles arise?

17. Is there a limit on how many alleles can exist for a gene?

18. How many genotypes would be possible in a population if there were 10 alleles for a gene?

19. In the fruit fly *Drosophila melanogaster*, eye color is normally dark red. Two recessive mutations called cinnabar and scarlet produce bright red eyes. How could one test whether the two mutations are alleles of the same gene, or variants of different genes?

20. Phenylketonuria (PKU) is a recessive disorder of amino acid metabolism in humans. The disease is characterized by an inability to convert the amino acid phenylalanine to the amino acid tyrosine. As a result phenylalanine is converted instead to phenylpyruvic acid, which is toxic. Individuals homozygous for the PKU allele show a range of symptoms. Why?

21. The disease PKU results in several symptoms, including light hair and abnormal brain development. Multiple traits resulting from a single gene is an example of _____.

22. Eye color in fruit flies is known to be affected by many genes. A recessive mutation at any one of these genes can, when in the homozygous condition, suppress the wild type eyes. What is this an example of?

23. Give an explanation for epistasis.

24. Coat color in mice is affected by at least five genes. The B gene determines the color of the pigment; B is black, b is brown. The C allele of the C gene permits color expression, but the recessive allele c is epistatic to the B gene. In a cross of *BbCc* × *BbCc*, predict the genotypic and phenotypic ratios.

25. In mice, the D gene controls the intensity of pigment specified by the other coat-color genes. *DD* and *Dd* permit normal expression of color, but dd results in a milky appearance. What is this an example of?

26. Predict the genotypic and phenotypic ratios from *BbDd* × *BbDd*.

27. In sweet peas, flower color (purple or albino) is affected by two genes C and P. In the following cross *CcPp* × *CcPp* of purple flowers, a 9:7 ratio of purple to white flowers is obtained. Explain this result.

28. Purple flowers in sweet peas are due to the pigment anthocyanin. Give an explanation for complementation in sweet peas in terms of anthocyanin synthesis.

29. Suppression of a gene's phenotype by another gene is always due to a recessive allele. True or false?

30. How many possible genotypes will there be for ten genes, six with two alleles and four with three alleles?

31. Many traits are quantitative, traits that are due to many genes interacting with the environment. Give two examples in people.

32. Explain why although the number of genotypes is discrete (countable) for a quantitative trait, quantitative traits generally show continuous variation.

33. What is the usual form of the distribution of trait values for a quantitative trait?

Check Yourself

1. Penetrance refers to whether or not an individual with a particular genotype shows the phenotype typically associated with that genotype. It is an all-or-none thing. When some individuals do not show a trait even though they have the appropriate genotype, the trait shows incomplete penetrance. Expressivity refers to the situation where there is a range of phenotypes for a particular genotype. **(Penetrance/expressivity)**

2. Incomplete penetrance. **(Penetrance/expressivity)**

3. Variable expressivity. **(Penetrance/expressivity)**

4. This ratio is the result of a monohybrid cross demonstrating incomplete dominance between the white and red alleles. **(Allelic interaction)**

5. $WW \times RW$ results in 1 WW: 1 RW and 1 white: 1 pink. **(Allelic interaction)**

6. That some form of allelic interaction other than dominance and recessiveness is at work. **(Allelic interaction)**

7. Incomplete dominance. Incomplete dominance does not mean that the phenotype for a heterozygote is exactly intermediate between the phenotypes of the homozygotes, only somewhere between those two phenotypes. **(Allelic interaction)**

8. Codominance. **(Allelic interaction)**

9. 1 $I^A I^A$: 1 $I^A I^B$: 1 $I^A I^O$: 1 $I^B I^O$, and 2 A: 1 B: 1 AB. **(Allelic interaction)**

10. Because AB individuals do not possess antibodies against A or B antigens. **(Allelic interaction)**

11. Because O individuals have red blood cells lacking antigens, which will not be attacked by any antibodies (of course, the antibodies in the blood of O individuals have to be first removed). **(Allelic interaction)**

12. Codominance. **(Allelic interaction)**

13. $c^+ > c^{ch} > c^h > c$. **(Allelic interaction)**

14. Recessive traits usually result from alleles that code for nonfunctional enzymes. An allele coding for a nonfunctional enzyme is recessive because if a dominant allele is present, the functioning enzyme the dominant allele codes for takes up the slack. **(Allelic interaction)**

15. Albinism is generally due to a recessive mutation that disables an enzyme needed for the synthesis of pigment. **(Allelic interaction)**

16. Mutations. **(Multiple allelism)**

17. No. In some natural populations dozens of alleles exist for a single gene. **(Multiple allelism)**

18. The formula is $n(n+1)/2$, where n is the number of alleles. Solving the formula gives 55. **(Multiple allelism)**

19. By the complementation test. If two individuals separately homozygous for the recessive mutations are crossed and the offspring are normal, the mutations exist at different genes. If, however, a mutant phenotype is produced, the mutations represent alleles of the same gene. (**Complementation test**)

20. Because of the environment. Individuals raised with larger quantities of phenylalanine in the diet will have more severe symptoms. (**Environment**)

21. Pleiotropy. (**Pleiotropy**)

22. A type of gene interaction called epistasis. (**Gene interaction**)

23. The inactivation of an enzyme catalyzing an intermediate step in a biochemical pathway. (**Gene interaction**)

24. 1 *BBCC*: 2 *BbCC*: 2 *BBCc*: 4*BbCc*: 2 *bbCc*: 1*bbCC*: 1 *BBcc*: 2 *Bbcc*: 1 *bbcc*, and 9 black: 3 brown: 4 albino. Note the non-Mendelian phenotypic ratio, a hallmark of gene interaction. (**Gene interaction**)

25. A type of gene interaction called modification. (**Gene interaction**)

26. 1 *BBDD*: 2 *BbDD*: 2 *BBDd*: 4 *BbDd*: 2 *bbDd*: 1 *bbDD*: 2 *Bbdd*: 1 *BBdd*: 1 *bbdd*, and 9 black: 3 dilute black: 3 brown: 1 dilute brown. (**Gene interaction**)

27. This is an example of the type of gene interaction called complementation. Complementation is epistasis at two or more genes. If either, or both genes are homozygous for the recessive allele, white flowers result. (**Gene interaction**)

28. Several enzymes are required for the synthesis of anthocyanin. If any one of the enzymes is disabled, anthocyanin will not be manufactured, and the flower will be white. (**Gene interaction**)

29. False. The phenomenon of suppression is the masking of a gene's phenotype by a dominant allele at another locus. (**Gene interaction**)

30. We can generalize the equation from question 18 for N loci: $n_1^{(n_1 + 1)}/_2 \times n_2^{(n_2 + 1)}/_2 \dots \times n_N^{(n_N + 1)}/_2$. Solving this equation gives the number of genotypes as 944,784. This shows that even for a small number of genes, the number of possible genotypes is enormous. (**Quantitative traits**)

31. Height, weight, skin color, IQ. (**Quantitative traits**)

32. Because each possible genotype will show a wide range of phenotype values if raised in different environments. (**Quantitative traits**)

33. Normal or bell-shaped curve. (**Quantitative traits**)

Grade Yourself

Circle the numbers of the questions you missed, then fill in the total incorrect for each topic. If you answered more than three questions incorrectly, you need to focus on that topic. (If a topic has less than three questions and you had at least one wrong, we suggest you study that topic also. Read your textbook, a review book, or ask your teacher for help.)

Subject: Extensions to Mendelian Genetics

Topic	Question Numbers	Number Incorrect
Penetrance/expressivity	1, 2, 3	
Allelic interaction	4, 5, 6, 7, 8, 9, 10, 11, 12, 13, 14, 15	
Multiple allelism	16, 17, 18	
Complementation test	19	
Environment	20	
Pleiotropy	21	
Gene interaction	22, 23, 24, 25, 26, 27, 28, 29	
Quantitative traits	30, 31, 32, 33	

Chromosomes

5

Brief Yourself

Chromosomes

Chromosomes were discovered in the second half of the nineteenth century by German cytologist W. Waldeyer. Chromosomes are extremely long, single molecules of DNA complexed with proteins. The combination of DNA and proteins is called chromatin. In procaryotes, chromosomes are relatively small and circular; in eucaryotes, chromosomes are relatively large and linear. After staining, mitotic chromosomes are easily visible under a light microscope. Work early in this century by the American geneticist Thomas Hunt Morgan established that genes reside on chromosomes. Thus, the abstract genetic "factors" posited to exist by Mendel were assigned to actual cell structures.

Describing Chromosomes

Chromosomes vary in several ways, including size, shape, number, and banding pattern. A karyotype is a photographic visualization of the set of chromosomes in a cell that allows chromosome description. Cytogenetics, the study of chromosomes, relies heavily upon karyotypes.

Chromosomal Sex Determination

In species showing separate sexes, sex is determined in a wide variety of ways. In many species, sex is determined by the chromosomes possessed by an individual. In a common mode that includes humans, sex is determined by a single pair of homologous chromosomes. In humans, females are XX and males are XY. In this system, the male is said to be heterogametic because he produces two different gametes relative to the sex chromosomes; hence, the male's gametic genotype determines the sex of the offspring. In other species, the situation is reversed with the female being the heterogametic sex. In this system, the female is said to be ZW and the male ZZ. In humans, the Y chromosome is small and lacks many of the genes present on the X chromosome (which provides the basis for X-linked sex linkage). One gene present on the Y that is missing on the X is the testis-determining factor (or sex-determining region Y). This gene appears to be a "master switch" that results in maleness.

Sex Linkage

A trait that is associated with one sex more often than the other sex is a sex-linked trait. Examples of sex-linked genetic diseases include hemophilia and red-green color blindness. Both of these are traits due to recessive alleles of genes on the X chromosome. Because the male possesses only one X chromosome, a male possessing a recessive allele will not have a corresponding normal allele to mask its effect. This is why males possess X-linked deleterious traits more frequently than females. To date, very few Y-linked traits have been identified, which is almost certainly due to the very few genes present on the Y chromosome.

 # Test Yourself

1. How many chromosomes in humans?

2. How many chromosomes in a human gamete?

3. How are the human chromosomes numbered?

4. Describe the shapes of the following chromosomes.

 (a)   (b)   (c)

5. Which pair of human chromosomes are the sex chromosomes?

6. Describe the shapes of the human sex chromosomes.

7. What are autosomes?

8. How does Down's syndrome arise?

9. Medically, how are karyotypes valuable?

10. What is a critical step in the actual preparation of a karyotype?

11. What are the types of rearrangements that occur to chromosomes during evolution?

12. Some eucaryotes are polyploid. Posit a mechanism by which a polyploid would arise from diploid parents.

13. Contrast aneuploidy and polyploidy.

14. In which group of organisms is polyploidy most common?

15. What are some characteristics of polyploid organisms compared to their diploid relatives?

16. Many agriculturally valuable polyploid plants are sterile. How are they propagated?

17. Why are many polyploids sterile?

18. Polyploids that result from hybridization between species are _____.

19. Polyploids that result from syngamy in a single species are _____.

20. The process whereby tissue-specific polyploidization is produced is _____.

21. Endomitosis occurs when _____ and _____occur, but are not followed by _____.

22. The polytene chromosomes of the salivary glands of *Drosophila* flies result from _____, without _____.

23. Provide an explanation of chromosomal bands.

24. There are many types of trisomy known in humans. What genetic diseases correspond to the following trisomies?

    47, XXY

    47, +21

25. Regions of a chromosome that are only mildly condensed are called what?

26. Regions of a chromosome that are maximally condensed are called what?

27. In humans, very few monosomic conditions are known. Why?

28. Are there any known monosomies in people?

29. How does synapsis occur when a chromosome of a homologous pair displays an inversion?

30. How does cri du chat syndrome in humans arise?

31. Give a plausible mechanism for the origin of duplications.

32. The X and Y chromosomes share no genes. True or False?

33. How would an XY human embryo develop if the testis-determining factor was inactivated?

34. What is a Barr body?

35. How do Barr bodies explain genetic mosaics in females (such as tortoiseshell cats)?

36. In humans, red-green color blindness is due to a recessive allele on the X chromosome. In a cross of a normal female that carries the defective allele and a normal male, predict the incidence of color blindness in the male and female offspring.

37. From question 36, what would be the probability of 3 normal and 2 colorblind offspring?

38. A female of some species with a phenotype due to a dominant allele *D* is mated with a male with the recessive phenotype due to *d*. All of the male offspring show the dominant phenotype, but none of the females do. Explain this result.

39. The following pedigree shows the transmission of a genetic disease. Is the mode of inheritance autosomal or sex-linked, dominant or recessive?

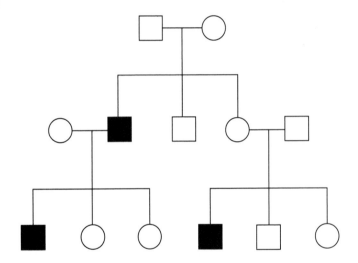

40. List the genotypes of the individuals from question 39.

41. Chromosomes are always condensed to some extent, especially during mitosis. What are the levels of condensation in eucaryotes?

42. What are the levels of chromosome condensation in procaryotes?

43. Why is chromosome condensation necessary?

Check Yourself

1. 46. (**Human karyotype**)

2. 23. (**Human karyotype**)

3. Human chromosomes, like the chromosomes of other species, are numbered by size from largest to smallest. In humans, there are two exceptions. By convention, the smallest pair are the 21st pair, the second smallest the 22nd pair. Also by convention, the sex chromosomes are the 23rd pair. (**Human karyotype**)

4. The position of the centromere determines one possible way to describe chromosomes. A. metacentric, B. acrocentric, C. telocentric. (**Chromosome description**)

5. The 23rd pair. (**Human karyotype**)

6. The X chromosome is a metacentric, the Y is an acrocentric. (**Human karyotype**)

7. Autosomes are the non-sex chromosomes. Hence, humans have 22 pairs of autosomes and 1 pair of sex chromosomes. (**Human karyotype**)

8. Down's syndrome is trisomy 21. One way it arises is from nondisjunction of the 21st chromosomes during anaphase I. This leads to a gamete carrying an extra 21st chromosome, which will produce a trisomic individual when it fuses to a normal gamete. (**Aneuploidy**)

9. They can reveal chromosomal rearrangements or aneuploidy, both of which cause genetic disease. (**Karyotypes**)

10. All the cells must be arrested at the start of mitosis. This is done by applying chemicals such as colchicine that inhibit the formation of the spindle apparatus. (**Karyotypes**)

11. Deletions, duplications, inversions, and translocations. (**Chromosome rearrangements**)

12. If one parent undergoes genome duplication in the right tissue, it will produce gametes with an extra set of chromosomes. Such a gamete will produce a polyploid individual after syngamy with a normal gamete. (**Polyploidy**)

13. Aneuploidy is the condition of having too many or too few of one particular kind of chromosome. Polyploidy is the condition of having extra sets of chromosomes. (**Polyploidy**)

14. Plants. One-half of all known plant genera have polyploid species. (**Polyploidy**)

15. Polyploid organisms tend to have larger cells relative to diploids. This characteristic can lead to greater yields in agriculturally valuable plants. (**Polyploidy**)

16. By asexual means such as grafting or cuttings. (**Polyploidy**)

17. Because of irregularities in meioisis leading to aneuploid gametes, which will produce inviable zygotes after syngamy. (**Polyploidy**)

18. Allopolyploids. **(Polyploidy)**

19. Autopolyploids. **(Polyploidy)**

20. Endomitosis. **(Polyploidy)**

21. Chromosome duplication and chromatid separation, cytokinesis. **(Polyploidy)**

22. Chromosome duplication, chromatid separation. **(Polytene chromosomes)**

23. Darkly staining bands correspond to tightly condensed DNA; light bands correspond to loosely condensed DNA. **(Chromosome description)**

24. 47, XXY = Klinefelter's syndrome; 47, + 21 = Down's syndrome. **(Aneuploidy)**

25. Euchromatin. **(Chromosome description)**

26. Heterochromatin. **(Chromosome description)**

27. Because monosomy is such a severe condition that zygotes with the condition fail to develop. **(Aneuploidy)**

28. Yes, Turner's syndrome, which is 45, X. In other words, individuals with this syndrome only have a single X chromosome. Phenotypically, these are sterile females, with webbed necks, hearing deficiencies, etc. **(Aneuploidy)**

29. In synapsis, homologous chromosomal regions pair up. When a chromosome carries an inversion, pairing will produce a loop in one of the two chromosomes, like the following: **(Chromosome rearrangements)**

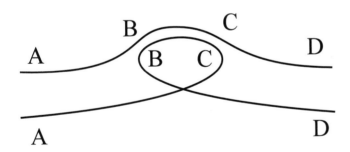

30. Cri du chat syndrome, characterized by severe mental retardation, is due to a deletion in the short arm of chromosome 5. The condition is named for the fact that people with this condition make a catlike sound. **(Chromosome rearrangements)**

31. The best theory currently is that most duplications are due to unequal crossing-over during meiosis. **(Chromosome rearrangements)**

32. False. The X and Y chromosomes share some genes called pseudoautosomal because they behave like autosomal genes. **(Sex chromosomes)**

33. The embryo would develop as a female because human embryos (and presumably other species with this type of sex determination) are genetically programmed to develop as females unless told otherwise by the testis-determining factor on the Y chromosome. The testis-determining factor presumably codes for a protein that acts as a "master-switch" to turn on a whole series of genes related to producing a male. (**Sex chromosomes**)

34. An inactivated X chromosome. In placental mammals, once the embryo has reached a few thousand cells, one of the X chromosomes in each cell randomly becomes inactivated; all cells descended from this cell will have the same X chromosome inactivated. The reason for the formation of Barr bodies is to compensate for the fact that males only have one X chromosome. (**Sex chromosomes**)

35. A genetic mosaic in females is the result of random X chromosome inactivation. Cells will only express alleles for genes on the X chromosome not condensed. Tortoise shell cats are females that show patches of black and orange fur. These are animals that are heterozygous for a gene on the X chromosome that affects fur pigment. The orange furs result from cells descended from cells that have inactivated the X chromosome carrying the black allele, and vice-versa for the orange furs. (**Sex chromosomes**)

36. None of the female offspring will be colorblind, but 1/2 of the males will be colorblind. (**Sex linkage**)

37. The probability of a colorblind child is 1/4. Thus, the probability of 3 normal and 2 colorblind children is $5!/3!2!(3/4)^3(1/4)^2 = 0.26$. (**Sex linkage**)

38. This would result if this were a sex-linked trait and the female was the heterogametic sex (ZW), while the male was the homogametic sex, as in many kinds of reptiles. (**Sex linkage**)

39. This pedigree is consistent with either an autosomal or X-linked recessive trait. However, because only males have the trait, it suggests that the trait is X-linked. (**Sex linkage**)

40. Denoting the normal allele as "n", and the disease-causing allele as "c", the genotypes are as follows. First generation: X^nY, X^nX^c; second generation: $X^nX^c, X^cY, X^nY, X^nX^c, X^nY$; third generation: $X^cY, X^nX^c, X^nX^c, X^cY, X^nY, X^nX^-$. (**Sex linkage**)

41. Chromosome condensation in eucaryotes involves three levels of organization: first, the DNA is wrapped around octamers of histone proteins (2H2A, 2H2B, 2 H3, 2H4). H1 histones serve as linkers to connect adjacent octamers. The combination of DNA plus the octamer and H1 forms a nucleosome. Nucleosomes are probably in direct juxtaposition along the length of the DNA molecule. Second, the nucleosomes are wound in a higher-order supercoil called the solenoid. In the third level of condensation, the solenoid is somehow organized upon a protein scaffold. The details of this third level of condensation are poorly understood. (**Chromosome condensation**)

42. In procaryotes, there are no histones. Chromosome condensation involves two levels: first, the DNA is negatively supercoiled, and second, the negatively coiled DNA is organized into 50–100 loops draped over a scaffold of protein and RNA. This is called the folded genome. (**Chromosome condensation**)

43. Two reasons. First, cells are too small to allow chromosomes to exist in a fully diffuse state. Second, condensed chromosomes are easily moved around during mitosis. (**Chromosome condensation**)

Grade Yourself

Circle the numbers of the questions you missed, then fill in the total incorrect for each topic. If you answered more than three questions incorrectly, you need to focus on that topic. (If a topic has less than three questions and you had at least one wrong, we suggest you study that topic also. Read your textbook, a review book, or ask your teacher for help.)

Subject: Chromosomes

Topic	Question Numbers	Number Incorrect
Human karyotype	1, 2, 3, 5, 6, 7	
Chromosome description	4, 23, 25, 26	
Aneuploidy	8, 24, 27, 28	
Karyotypes	9, 10	
Chromosome rearrangements	11, 29, 30, 31	
Polyploidy	12, 13, 14, 15, 16, 17, 18, 19, 20, 21	
Polytene chromosomes	22	
Sex chromosomes	32, 33, 34, 35	
Sex linkage	36, 37, 38, 39, 40	
Chromosome condensation	41, 42, 43	

Linkage

6

Brief Yourself

Linkage

Mendel's Law of Independent Assortment states that alleles at two or more loci will assort independently into gametes. To express this precisely, every possible gametic genotype for two or more genes is equally likely. The number of possible gametes an individual can produce for some number of genes where n of them are heterozygous is 2^n. Thus, for an individual with the genotype *AaBbCc*, there will be $2^3 = 8$ possible gametic genotypes. Under independent assortment, each of them will be equally likely (p = ⅛).

Independent assortment results when genes are on different chromosomes or far apart on the same chromosome. Genes that co-occur on the same chromosome are linked; genes that occur on different chromosomes are unlinked. Alleles for linked genes will tend to assort together into gametes. But is linkage absolute? We might expect that an individual who carries *A* and *B* alleles on one chromosome, and *a* and *b* on the homologous chromosome will only produce *AB* or *ab* gametes. However, if crossing-over occurs between those two genes during meiosis, *Ab* and *aB* gametes will also be produced, although generally in lesser numbers than *AB* or *ab*. The *AB* and *ab* gametes are called the parental types; the *aB* and *Ab* gametes are the recombinant types. The summed frequency of the two classes of recombinant gametes is the recombination frequency and is denoted r_{AB}. For example, if the frequency of *aB* = .20 and the frequency of *Ab* = .20, then r_{AB} = .40.

Gene Mapping Using the Frequency of Recombination

In 1911, Alfred Sturtevant, a student of the great geneticist Thomas Hunt Morgan, realized that the frequency of the recombinant gametes reflects the distance apart two linked genes are on a chromosome. The logic is very simple: the further apart two linked genes are, the greater the probability that crossing-over will occur between them, and thus the greater the frequency of recombinant gametes. He realized that if data on recombinant frequencies were available for several linked genes, the genes could be mapped on the chromosome. Linkage mapping accomplishes two goals:

- establishing the relative order of the genes, and

- establishing the relative distances of the genes from each other.

Crossing-Over

Linkage mapping depends on the phenomenon of crossing-over, which is the process during meiosis (prophase I) where paired, homologous chromosomes exchange segments of DNA. Under a microscope, crossing-over manifests itself visibly as chiasmata (chiasma, s). One or multiple cross-overs can occur, and any combination of the four chromatids (the two duplicated homologous chromosomes) can be involved (as long as it is only two at a time per chiasma). For example, a double cross-over might involve the same two non-sister chromatids, or it might involve all four chromatids. Each possible number of cross-over events (e.g., a single cross-over, a double cross-over, etc.) will produce an average of 2 recombinant gametes out of a total of four meiotic products considering all the possible combinations of strands that can be involved in a single cross-over, or a double cross-over, etc. Thus, the maximum value for the recombination frequency for two linked genes is 50%, as one would intuitively expect.

 # Test Yourself

1. Enumerate every possible gametic genotype from an individual with the genotype *AaBbCc*.

2. Enumerate every possible gametic genotype from an individual with the genotype *AaBbCc* if the A and C genes are completely linked (no crossing-over). Assume that the parents were *AABBCC* and *aabbcc*.

3. How many gametes are possible from an individual with the genotype *AaBbCcDd* if the A and B genes are completely linked, and the C and D genes are completely linked?

4. How many gametes are possible from an individual with the genotype *AaBbCcDd* if all four genes are unlinked?

5. Consider an offspring from the cross *AABB* × *aabb*. What kinds of gametes will it produce under complete linkage?

6. Consider an offspring from the cross *AABB* × *aabb*. What kinds of gametes will it produce and in what proportion if $r_{AB} = 40\%$?

7. Predict the products of the following cross-over configuration for a pair of homologous chromosomes in meiosis.

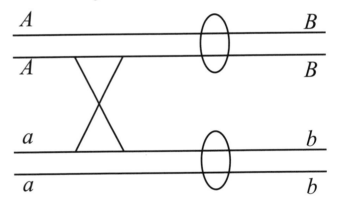

8. Predict the products of the following cross-over configuration for a pair of homologous chromosomes in meiosis.

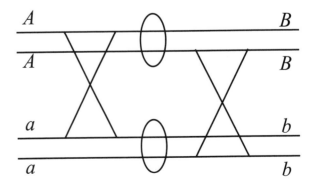

9. Predict the products of the following cross-over configuration for a pair of homologous chromosomes in meiosis.

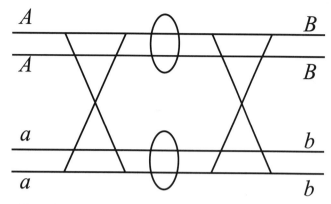

10. If two offspring from question 6 were crossed, what would be the probability of an *AaBb*?

11. If two offspring from question 6 were crossed, what would be the probability of an *AaBb* followed by a *AaBB*?

12. If *AaBb* is crossed to *aaBB* and there is complete linkage, predict the genotypic ratio of the offspring. Assume the parents of the first individual to be *AAbb* and *aaBB*.

13. From the cross *AaBb** × *aabb*, the following genotypic ratio of the offspring is observed: 3*Aabb*: 3*aaBb*: 2*AaBb*: 2*aabb*. What were the gametic genotypes that formed the individual *AaBb**?

14. From question 13, what is the importance of the *aabb* individual from the original cross?

15. From the cross *AaBbCc* × *aabbcc*, 40% of the offspring are *AaBbcc*. Are the genes linked or not?

16. From the cross *AaBbCc* × *aabbcc*, 126 out of 1000 offspring are *AaBbcc*. Are the genes linked or not?

17. An individual with the genotype *AaBb* produces the following gametes out of 1000: 246 *AB*, 251 *ab*, 247 *Ab*, 256 *aB*. Perform a chi-square test of the hypothesis that there is no linkage.

18. From the cross *AaBbCc* × *aabbcc*, 1% of the offspring are *AabbCc*. Explain.

19. What would be the probability of choosing seven genes in humans all on different chromosomes? Assume the same number of genes on each chromosome.

20. What would be the probability of choosing two genes in humans on the same chromosome? Assume the same number of genes on each chromosome.

21. The cross *AABB* × *aabb* yields offspring *AaBb*, which produce gametes in the following proportion: *AB* = .35, *Ab* = .15, *ab* = .35, *aB* = .15. What is the map distance between the A and B genes?

22. What is the maximum recombination frequency between two linked genes?

23. The cross *AABB* × *aabb* yields offspring *AaBb*, which produce gametes in the following proportion: *AB* = .15, *Ab* = .35, *ab* = .15, *aB* = .35. Is this possible?

24. How many possible gene orders on a eucaryotic chromosome are there for three genes? Enumerate them.

25. How many possible gene orders on a eucaryotic chromosome are there for ten genes?

26. How many possible gene orders on a procaryotic chromosome are there for three genes? Enumerate them.

27. How many possible gene orders on a procaryotic chromosome are there for ten genes?

28. From two-point crossing experiments, the following data are obtained:

r_{AB} = .29

r_{AC} = .17

r_{CB} = .12

Draw the linkage map and give the distances.

29. From two-point crossing experiments, the following data are obtained:

$r_{DC} = .25$

$r_{AC} = .40$

$r_{AB} = .49$

$r_{CB} = .30$

$r_{DB} = .48$

$r_{AD} = .15$

Draw the linkage map and give the distances.

30. Consider the following cross: *AAbbcc × aaBBCC*. If an offspring is crossed with a tester (*AaBbCc × aabbcc*), the following proportions of offspring are observed:

Aabbcc	510
aaBbCc	505
AaBbCc	45
aabbcc	50
AabbCc	80
aaBbcc	85
AaBbcc	3
aabbCc	5.

Draw the linkage map and the distances.

31. Calculate the interference from question 30.

32. The following cross is performed: *Aabbcc × aaBBCC*.

What gametic classes will an offspring from this cross produce and in what proportions if there is 30% interference and the linkage map is
A——.15——B————.20————C?

33. The following cross is performed: *AaBbCcDd × aabbccdd*. The offspring are found to occur in the following numbers out of 1000 offspring:

aaBbCcDd	42
Aabbccdd	43
AaBbCcdd	140
aabbccDd	145
aaBbccDd	6
AabbCcdd	9
AaBbccdd	305
aabbCcDd	310

Which genes, if any, are linked?

34. Draw a linkage map for the genes in question 33.

35. What are the advantages of three-point crosses over two-point crosses for gene mapping?

36. There is a direct proportionality (linear relationship) between r values and actual physical distance along a chromosome. True or false?

37. What does a m.u. of 1 mean?

38. What would be the recombination frequency if every single meiotic cell experienced crossing-over between the two genes under consideration?

39. Do sister-chromatid cross-overs occur?

Check Yourself

1. *ABC, ABc, Abc, abc, AbC, aBc, aBC, abC.* (**Gametic genotypes**)

2. *ABC, abc, AbC, aBc.* (**Gametic genotypes**)

3. This is equivalent to assuming that there are two fewer genes. Thus, the number of gametes will be $2^2 = 4$. (**Gametic genotypes**)

4. $2^4 = 16$. (**Gametic genotypes**)

5. *AB* and *ab*. (**Gametic genotypes**)

6. The frequency of the recombinants *Ab + aB* = 40%; hence, the frequency of *AB + ab* = 60%. Thus, the proportions are: 30% *AB*, 30% *ab*, 20% *Ab*, 20% *aB*. (**Gametic genotypes**)

7. From top to bottom: *AB, aB, Ab, ab*. (**Crossing-over**)

8. From top to bottom: *aB, Ab, Ab, aB*. (**Crossing-over**)

9. From top to bottom: *ab, AB, ab, AB*. (**Crossing-over**)

10. This problem can be solved by working through a Punnett square. However, this time it is necessary to write the frequencies of the gametes in front of the gametes and then multiply those coefficients throughout (this was not done before with simple Mendelian crosses because every gamete was assumed to occur in equal proportion). The Punnett square would be set up as follows:

	.30 *AB*	.30 *ab*	.20 *Ab*	.20 *aB*
.30 *AB*				
.30 *ab*				
.20 *Ab*				
.20 *aB*				

    In the interest of saving time, one does not need to fill it in completely, only the part that corresponds to the outcome *AaBb*:

	.30 *AB*	.30 *ab*	.20 *Ab*	.20 *aB*
.30 *AB*		.09		
.30 *ab*	.09			
.20 *Ab*				.04
.20 *aB*			.04	

    Summing these numbers together gives us the desired probability: .09 + .09 + .04 + .04 = .26. (**Dihybrid cross with linkage**)

11. The probability (using the same method as for question 10) of an *AaBB* is .12. Thus, the desired probability is .26 × .12 = .031. (**Dihybrid cross with linkage**)

12. If there is complete linkage, then the only possible gametes from the first individual are *Ab* and *aB*. Thus, the Punnett Square is set up as:

	Ab	*aB*
aB	*AaBb*	*aaBB*

Thus, *AaBB* and *aaBB* occur in equal proportions. (**Dihybrid cross with linkage**)

13. The most numerous classes of offspring reflect the syngamy of nonrecombinant gametes, which reflects the original gametes that fused to make individual 1. In this case, the gametes were *Ab* and *aB*. (**Gametic genotypes**)

14. The *aabb* is a tester, an individual showing no variation for the genes under consideration. The tester allows one to classify each offspring resulting from a test cross according to the gametes that it received from the non-tester parent. This is done by simply subtracting from the offspring the genetic contribution from the tester (in this case, *ab*). What is left over reflects the gametic genotype contributed from the other parent. (**Gametic genotypes**)

15. The genes are linked because if they were unlinked, all 8 possible gametes would occur in equal proportions. (**Linkage**)

16. Probably unlinked because 126 out of 1000 is very close to $\frac{1}{8}$ = .125. (**Linkage**)

17. A chi-square test is one of the most important tools in classical genetics. Briefly, it is a goodness of fit test that tells us whether observed numbers are consistent with an hypothesis we may hold. The statistic is calculated as $\Sigma\,(O - E)^2/E$, where O = the observed value, E = the expected value. In this case, E = 1/4(1000) = 250. Doing the arithmetic: $(246 - 250)^2/250 + (251 - 250)^2/250 + (247 - 250)^2/250 + (256 - 250)^2/250 = 62/250 = .25$. With $4 - 1 = 3$ degrees of freedom, the value of .25 is well below the critical (at a probability of 5%) value of 7.8. Thus, we accept the hypothesis that there is no linkage. (**Linkage**)

18. The only way to produce an *AabbCc* offspring would be if the first individual produced an *AbC* gamete (remember the function of the tester). This type of gamete was likely produced by a double cross-over as shown by the low probability. (**Crossing-over**)

19. $46!/(46 - 7)! \times (1/46)^7 = .62$. The second term is the probability of picking seven prespecified chromosomes in a row out of 46. This is multiplied by the first term, which is the total number of ways to pick seven chromosomes without repeat out of 46 chromosomes. (**Probability of linkage/no linkage**)

20. $46/46^2 = 1/46 = .0217$. This problem is solved with the same logic as the prior one. There are 46^2 ways to choose 2 chromosomes out of 46; the numerator counts the subset of these ways corresponding to the same chromosome picked twice. (**Probability of linkage/no linkage**)

21. The map distance m.u. (or cM = centiMorgans) = the recombinant frequency × 100 = $r_{AB} \times 100 = 30$. (**Linkage mapping**)

22. .50. This value should be intuitive. However, proof of it is mathematically involved. (**Linkage mapping**)

23. Sure, if we assume that *A* and *b,* and *a* and *B* were linked together on the same chromosomes. The student should avoid the mindset that in linkage problems *A* and *B*, and *a* and *b* always go together on the same chromosome. **(Gametic genotypes)**

24. The number of genes orders on a linear chromosome is $n!/2$, where n is the number of genes. For three genes, there are $3!/2 = 3$ orders, which are ABC, BCA, and CAB. The student should understand that ABC and CBA are identical gene orders. Turning a gene order around does not make it a new one. **(Linkage mapping)**

25. $10!/2 = 1,814,400.$ **(Linkage mapping)**

26. On a circular chromosome, the number of gene orders is $(n-1)!/2$. For three genes, there is only one gene order:

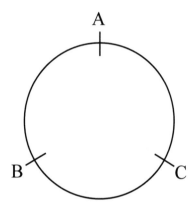

The student should verify that there are no other gene orders in this example. **(Linkage mapping)**

27. $(10-1)!/2 = 181,440.$ **(Linkage mapping)**

28. The largest distance immediately tells us that the A and B genes are farthest apart, with C in the middle. drawing the map so far:

A   C   B

Next, we need to add the distances:

A   17   C   12   B. **(Linkage mapping)**

29. Start with any three genes and do as in question 28. Then pick the last gene and find where it fits, leading to the map:

A   15   D   25   C   30   B.

Notice that some of the distances come close but do not exceed .50. It is possible for very far apart genes on a single chromosome to approach a recombination frequency of .50. **(Linkage mapping)**

30. This is data resulting from a three-point cross, where the offspring are classified according to three loci, not just two. To solve this problem, first cross off the contribution from the tester *abc* to each offspring type, leaving behind the genetic contribution from the *AaBbCc* individual:

Abc	510
aBC	505
ABC	45
abc	50
AbC	80
aBc	85
ABc	3
abC	5
	1283

The secret to a three-point cross is to find the gamete classes representing double recombination events. Because of the rarity of double recombination, the classes represented by 3 and 5 individuals are the double recombinants. These are compared to the chromosomal constitution of the parent; the gene that was transposed to form the double recombinant classes is the gene in the middle. This places the B gene in the middle:

A   B   C

Now we need to calculate the distances. To do this, we need to find the recombinant frequencies between each pair of genes. First, one should recognize that the classes *ABC* and *abc* represent single cross-over events between *a* and *b*, remembering that the genetic constitution of the parent is *Abc/aBC*; but, the double recombinants also represent recombination between *a* and *b*. Thus, $r_{AB} = {}^{(45 + 50 + 3 + 5)}/_{1283} = .08$. And $r_{BC} = {}^{(80 + 85 + 3 + 5)}/_{1283} = .135$. The final map is:

A   8   B   (13.5)   C **(Linkage mapping)**

31. Interference (I) is a measure of the extent to which crossing-over events for 3 genes are non-independent. If I = 100%, then there is complete nonindependence, meaning that if a cross-over occurs between genes A and B, it prevents a second cross-over from occurring between B and C. If I = 0%, the cross-overs are completely independent. The formula is I = 1 − (observed frequency of double cross-overs/expected frequency of double-crossovers). The expected frequency is calculated under the assumption of complete independence, in other words as the probability of a single cross-over between A and B multiplied by the probability of a single cross-over between B and C. In the present context, the probabilities of single cross-overs are estimated by the map distances. Hence, the expected probability of the double cross-over is $.08 \times .135 = .0108$. And, I = 1 − (.0062/.0108) = .42, or in other words an interference of 42%. **(Interference)**

32. This is a difficult problem. First, one should use the formula for I to calculate the observed frequency of double recombination: $.30 = 1 - (obs/.03)$, $obs = .0210$. Using this information and the information from the map on the single cross-over probabilities, we can calculate the expected gametic frequencies. First, the nonrecombinants are Abc and aBC and their frequencies can be calculated as:

$$Abc = 1/2[1 - r_{AB} - r_{BC} + P(\text{Double cross-over})] = .336$$

$$aBC = 1/2[1 - r_{AB} - r_{BC} + P(\text{Double cross-over})] = .336.$$

The 1/2 is because there are two nonrecombinant gametes. The probability of double recombination has to be added back because by subtracting off r_{AB} and r_{BC}, we are subtracting off the frequency of double recombination twice (remember that r_{AB}, r_{BC} were calculated by the addition of the frequency of double recombination). Using this logic, we calculate the remaining frequencies:

$$ABC = 1/2(r_{AB} - .021) = .065$$

$$abc \quad = 1/2(r_{AB} - .021) = .065$$

$$AbC \ = 1/2(r_{BC} - .021) = .090$$

$$aBc \ = 1/2(r_{BC} - .021) = .090$$

$$Abc = 1/2(.021) = .010$$

$$abC = 1/2(.021) = .010. \textbf{ (Interference)}$$

33. First, we know there must be complete linkage between a single pair of genes because there are only 8 instead of $2^4 = 16$ gametic classes (after elimination of the abc contribution from the tester). To learn which genes are linked, simply calculate the r value for pairs of genes. If it is less than .50, the genes are linked.

    A and B genes

    $AB = 140 + 305 = 445$

    $ab = 145 + 310 = 455$

    $Ab = 43 + 9 = 52$

    $aB = 42 + 6 = 48$

Clearly the genes are linked and are 10 (= 52 + 48/1000) m.u. apart on the same chromosome.

    A and D genes

    $AD = 0$

    $ad = 0$

    $Ad = 43 + 140 + 9 + 305 = 497$

    $aD = 42 + 145 + 6 + 310 = 503$

Clearly the genes are completely linked and separated by 0 m.u.

    C and D genes

    $CD = 42 + 310 = 352$

    $cd = 43 + 305 = 348$

    $Cd = 140 + 9 = 149$

    $cD = 145 + 6 = 151$

The two genes are 30 m.u. apart on the same chromosome.

B and C genes

$BC = 42 + 140 = 182$

$bc = 43 + 145 = 188$

$Bc = 6 + 305 = 311$

$bC = 9 + 310 = 319$

The two genes are 37 m.u. apart on the same chromosome. Hence, all the genes are linked. **(Linkage)**

34. Using the m.u. data from question 33 yields the linkage map:

<u>B   10   A/D   30   </u> C. **(Linkage mapping)**

35. There are two advantages in three-point crosses over two-point crosses. First, data collection is more efficient because one only has to perform a single three-point cross to map three genes. With two-point crosses it is necessary to do three crosses to map three genes. Second, one gains data on the interference among cross-over events. **(Linkage mapping)**

36. False. If r values accurately reflected the average number of cross-over events between two genes, then r values would be directly proportional to linear distance along a chromosome. However, the further apart two genes are, the greater will be the number of multiple cross-overs, which re-create the parental type gametes and hence cause the actual map distance to be underestimated. **(Linkage mapping)**

37. A m.u. of 1 means one recombinant gamete per 100 meiotic products. **(Linkage mapping)**

38. As explained in "Brief Yourself," the maximum value for r is .50. **(Linkage mapping)**

39. Yes, but they cannot be detected because sister-chromatids are identical. **(Crossing-over)**

Grade Yourself

Circle the numbers of the questions you missed, then fill in the total incorrect for each topic. If you answered more than three questions incorrectly, you need to focus on that topic. (If a topic has less than three questions and you had at least one wrong, we suggest you study that topic also. Read your textbook, a review book, or ask your teacher for help.)

Subject: Linkage

Topic	Question Numbers	Number Incorrect
Gametic genotypes	1, 2, 3, 4, 5, 6, 13, 14, 23	
Crossing-over	7, 8, 9, 18, 39	
Dihybrid cross with linkage	10, 11, 12	
Linkage	15, 16, 17, 33	
Probability of linkage/no linkage	19, 20	
Linkage mapping	21, 22, 24, 25, 26, 27, 28, 29, 30, 34, 35, 36, 37, 38	
Interference	31, 32	

Advanced Linkage Analysis

7

Brief Yourself

The previous chapter dealt with the topic of linkage and how the frequency of recombinant gametes can be used to map genes on a chromosome. Typically, textbooks on genetics include an additional chapter on linkage that expands on the subject by discussing ways to correctly map distances, linkage analyses based on the products of single meioses, and human linkage analysis. The correction of map distances is important because the recombination frequency for two genes will underestimate the true physical distance, especially for distances above 25 m.u. This is because the farther two genes are apart, the greater the chance of multiple cross-overs which can regenerate nonrecombinant gametes. This results in the maximum value for r being .50. Ideally, we would like to be able make the map distance between two genes equal to the average number of cross-overs (= chiasmata) occurring between the two genes because this quantity is directly proportional to physical distance along a chromosome.

Test Yourself

1. Ideally, what would map distances on a chromosome represent?

2. What is perhaps the simplest way to approximate a direct proportionality (linear relationship) between map distance and actual physical distance between pairs of genes along a chromosome?

3. On the following linkage map, what would be an improved estimate of the distance between A and D?

   A   .20   B   .17   C   .22   D

4. Is there a constraint on the magnitude of map distances calculated by summing all intervening distances?

5. In the absence of a series of intervening r values, what can we do to arrive at a corrected estimate for the distance between two widely separated genes?

6. What would be the corrected distance for r = .10?

7. What would be the corrected distance for r = .30?

8. Some eucaryotic species produce structures that allow all four products of a single meiosis to be studied. Give examples of these species.

9. What are the general details of the Ascomycete life cycle?

10. What is the difference between an ordered and unordered tetrad (or octad)?

11. Yeast (*Saccharomyces cerevesiae*) produce an ascus that represents an unordered tetrad. For a cross *AB* × *ab* (remember, the cells that fuse to form the zygote are haploid), draw the possible tetrads if the genes are unlinked. Label the spores by genotype.

12. Considering yeast, for a cross *AB* × *ab* (remember, the cells that fuse to form the zygote are haploid), draw the possible tetrads if the genes are completely linked. Label the spores by genotype.

13. Considering yeast, for a cross *AB* × *ab* (remember, the cells that fuse to form the zygote are haploid), draw the possible tetrads if the genes are linked, but recombination does occur. Label the spores by genotype.

14. How could we use data on the frequency of the three yeast tetrad types to determine if two genes are linked or not?

15. The following data is obtained on the occurrence of yeast tetrads based on two genes: PD = 51, NPD = 60, T = 2. Use a chi-square test to test the hypothesis that there is no linkage.

16. Assuming two yeast genes are unlinked, how is a tetratype tetrad produced?

17. The following data is obtained on the occurrence of yeast tetrads based on two genes: PD = 51, NPD = 4, T = 20. Use a chi-square test to test the hypothesis that there is no linkage.

18. From question 17, calculate the map distance between the two genes.

19. Calculate a corrected map distance from question 17.

20. What is a type of linkage mapping that can be performed with ordered tetrads or octads?

21. For a single genetic marker in *Neurospora crassa*, a mold producing an ordered octad, what are the two patterns of octads with regard to the genotypes of the spores?

22. How is a second division segregation octad produced in *Neurospora*?

23. In a classic experiment, 132 asci were analyzed in *Neurospora* for the thi locus. Of these, 28 were found to be of the second division segregation pattern. How far is the thi locus from the centromere?

24. In humans, it is difficult to perform classic linkage analyses because of the impossibility of obtaining large numbers of offspring from a single mating. What are some alternative techniques for determining if a pair of genes are linked?

25. What are some techniques in humans for locating genes to particular chromosomes?

26. The following human pedigree traces the transmission of two genetic diseases. The first is due to a dominant allele *A* and individuals with the allele are shown in dark. The second disease is due to a recessive allele *b* and individuals with the condition are shown with an asterisk. The genotypes of the original parents are shown. Are these genes linked?

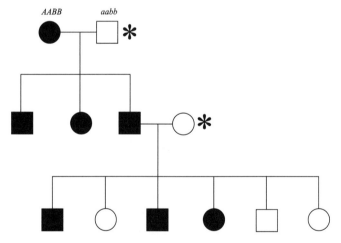

27. If the frequency of recombination between the genes from question 26 is .20, what is the probability of a individual with only the second disease coming from the original cross (generation 1)?

28. What is the goal of the Human Genome Project (HGP)?

29. How many human genes have been identified so far by HGP?

Check Yourself

1. The average number of cross-overs between a pair of genes. (**Map distance correction**)

2. Simply estimate the map distance between two genes as the sum of all the intervening map distances. (**Map distance correction**)

3. .20 + .17 + .22 = .59. (**Map distance correction**)

4. No. Map distances between a pair of genes calculated by summing all intervening distances can exceed .50. The longer the chromosome, the greater the distances. (**Map distance correction**)

5. We could correct the r value if we had a mathematical equation that related the recombination frequency to the average number of cross-overs. In fact, the equation is:

   $$m = -\ln(.5 - r) - .6931,$$

   where m is the average number of cross-overs occurring between two genes. If we plug in the observed value of r, we can then solve for m. The quantity m could serve as a measure of the distance between two genes. However, if we want to use the same scale as r, we can divide m by 2. This gives us a corrected value for r that is no longer constrained to be less than or equal to .50. (**Map distance correction**)

6. $m = -\ln(.5 - .1) - .6931 = .22$; $.22/2 = .11$. (**Map distance correction**)

7. $m = -\ln(.5 - .3) - .6931 = .92$; $.92/2 = .46$. Notice that the corrected estimate for .10 is .11, which is very close to .10, but that the corrected value for .30 is .46. Higher values of r require greater amounts of correction. (**Map distance correction**)

8. Some single-celled algae and fungi of the class ascomycetes, which includes yeast. (**Analysis of single meioses**)

9. Fungi of the class Ascomycetes have a haploid life cycle, meaning that the normal condition of the cells is to be haploid. Sexual reproduction occurs when cells of opposite mating strains fuse to form a diploid zygote, which quickly undergoes meiosis to produce four haploid cells. The new haploid cells, which will develop into spores, are contained together in a structure called an ascus, a feature that allows us to study all four products of a single meiosis. The four meiotic products contained within the ascus are a tetrad. In some Ascomycetes, the four meiotic products go through a round of mitosis resulting in an octad. (**Analysis of single meioses**)

10. In an unordered tetrad or octad, the array of spores does not reflect in any way the positional relationships of the cells as they were produced by the divisions of meiosis. In an ordered tetrad or octad, the spores exist in a linear array constrained by the structure of the ascus. This results in a preservation of the positional ordering of the cells. (**Analysis of single meioses**)

11. There are three possible tetrads considering the array of genotypes of the spores: the parental ditype ("ditype" because there are two genotypes) (PD), nonparental ditype (NPD), and tetratype (T). These can be diagrammed as follows:

PD	NPD	T
AB	*Ab*	*Ab*
AB	*Ab*	*AB*
ab	*aB*	*aB*
ab	*aB*	*ab*

    Note that the tetrads *AB*, *AB*, *ab*, *ab* and *AB*, *ab*, *AB*, *ab* are considered the same. (**Analysis of single meioses**)

12. If the genes are completely linked, only PDs can arise. (**Analysis of single meioses**)

13. If recombination does occur, then all three kinds of tetrads (PD, NPD, T) are possible, just as when there is no linkage. (**Analysis of single meioses**)

14. The key is to realize that if the genes are unlinked, then #PDs = #NPDs; if, on the other hand, the genes are linked, then PDs > NPDs. (**Analysis of single meioses**)

15. Under the hypothesis that there is no linkage, then #PDs = #NPDs. The expected frequency for each class is $(51 + 60)/2 = 55.5$. Doing the arithmetic: $(51 - 55.5)^2/55.5 + (60 - 55.5)^2/55.5 = .73$. The value of .73 for 1 $(= 2 - 1)$ degree of freedom is less than the 5% critical value of 3.8. Thus, we accept our hypothesis that the genes are unlinked. (**Analysis of single meioses**)

16. By a single cross-over involving non-sister chromatids between a gene and its centromere. (**Analysis of single meioses**)

17. Under the hypothesis that there is no linkage, then #PDs = #NPDs. The expected frequency for each class is $(51 + 4)/2 = 27.5$. Doing the arithmetic: $(51 - 27.5)^2/27.5 + (4 - 27.5)^2/27.5 = 40.2$. The value of 40.2 for 1 $(= 2 - 1)$ degree of freedom is much greater than the 5% critical value of 3.8. Thus, we reject our hypothesis that the genes are unlinked in favor of the alternative hypothesis that the genes are linked. (**Analysis of single meioses**)

18. We can calculate map distance as the recombinant frequency r. In this case $r = (\frac{1}{2}T + NPD)/(PD + NPS + T)$. This is because half of the spores (= gametes) produced by a T tetrad will be recombinant, but all of the spores from a NPD are recombinant. Doing the arithmetic: $(\frac{1}{2} \times 20 + 4)/75 = .19$. Thus, the genes are 19 m.u. apart on the same chromosome. (**Analysis of single meioses**)

19. We can arrive at an improved estimate of map distance by realizing that the average number of cross-overs $m = (T + 6NPD)/(PD + NPD + T)$. Doing the arithmetic gives $m = .59$. When divided by 2, m gives a corrected map distance; in this case, $m/2 = .29$. (**Analysis of single meioses**)

20. Centromere mapping, the placement of a gene relative to the centromere. (**Analysis of single meioses**)

21. The first division segregation pattern, where all the *A* alleles are at one end, and all the *a* alleles are at the other end. Or the second division segregation pattern, where there is an alternating pattern of A and a alleles. (**Analysis of single meioses**)

22. By crossing-over that occurs between a gene and the centromere. (**Analysis of single meioses**)

23. One-half of the second division segregation octad's spores correspond to recombinant spores; hence the r value between thi and the centromere is $(\frac{1}{2}) \times (^{28}/_{132}) = .11$m.u. (**Analysis of single meioses**)

24. Pedigree analysis, somatic cell hybridization, and deletion or duplication mapping. (**Human linkage analysis**)

25. Somatic cell hybridization, and deletion or duplication mapping. (**Human linkage analysis**)

26. Clearly the genes are linked because the dominant A and B alleles are inherited together, as revealed by the third generation. (**Human linkage analysis**)

27. 10%. (**Human linkage analysis**)

28. To map and sequence the entire human genome. (**Human linkage analysis**)

29. Approximately 5,000 of the estimated 50,000–100,000 human genes have been identified by HGP. The progress of the HGP can be followed on their Web site (www.nhgri.nih.gov/HGP/). (**Human linkage analysis**)

Grade Yourself

Circle the numbers of the questions you missed, then fill in the total incorrect for each topic. If you answered more than three questions incorrectly, you need to focus on that topic. (If a topic has less than three questions and you had at least one wrong, we suggest you study that topic also. Read your textbook, a review book, or ask your teacher for help.)

Subject: Advanced Linkage Analysis

Topic	Question Numbers	Number Incorrect
Map distance correction	1, 2, 3, 4, 5, 6, 7	
Analysis of single meioses	8, 9, 10, 11, 12, 13, 14, 15, 16, 17, 18, 19, 20, 21, 22, 23	
Human linkage analysis	24, 25, 26, 27, 28, 29	

Quantitative Genetics

8

Brief Yourself

Quantitative Traits

As seen in Chapter 4 (Extensions to Mendelian Genetics), most traits are not simple Mendelian traits, controlled by a single gene with dominant and recessive alleles determining discrete trait categories (e.g., purple flowers vs white flowers). Probably most traits are controlled by multiple genes. A quantitative trait is one that is controlled by many genes, each making a small contribution to the phenotype, as well as by the environment. Polygenic traits show continuous variation (however, some polygenic traits show discrete variation), which means that there are potentially an infinite number of trait values. In humans, examples of quantitative traits include weight, height and skin color. Because the genes controlling quantitative traits have not been identified, we must study quantitative traits statistically, rather than in a more exact, analytical fashion.

Basic Statistics

The mean (or average) is calculated as follows: $\overline{X} = \frac{1}{n} \Sigma\, X_i$.

The symbol Σ means "to take the sum of". For example, if $X_i = 1, 3, 4$, then $\Sigma\, X_i = 1 + 3 + 4 = 8$. The variance measures the variation among a set of numbers and is calculated as follows:

$$s^2 = \frac{1}{(n-1)} \Sigma\, (X_i - \overline{X})^2 \,.$$

The standard deviation is simply the square root of the variance: $s = \sqrt{s^2}$.

Often, we want to measure the association between two variables X and Y; this is done with the correlation coefficient: $\dfrac{1}{(n-1)S_x S_y} \Sigma\, (X_i - \overline{X})(Y_i - \overline{Y})$.

The quantity $\Sigma\, (X_i - \overline{X})(Y_i - \overline{Y})/(n-1)$ is called the covariance and is symbolized as cov_{XY}.

Heritability

Early in this century, R. A. Fisher, British mathematician and statistician, pioneered heritability analysis, the statistical method of analyzing quantitative traits. Heritability is based on the concept of partitioning the variance for a phenotypic trait into genetic and environmental components. It relies on the simple statistical fact that if a number represents the sum of two other numbers (for example, a household income would be the sum of the husband's and wife's incomes), the variance (using V instead of s^2) of that number can be partitioned as follows: $V_Z = V_X + V_Y + 2 \, cov_{XY}$.

If we consider the value for a quantitative trait in a particular individual to represent the sum of genetic and environmental contributions, then we can write: $VP = V_G + V_E + 2 \, cov_{GE}$, where V_P is the variance of the phenotype, V_G is the genetic variance, and V_E is the environmental variance. If we assume that there is no interaction between the genes and the environment, then the equation reduces to: $V_P = V_G + V_E$.

Fisher defined the broad-sense heritability H^2 of a trait to be the proportion of the phenotypic variance due to the genetic variance: $H^2 = V_G / V_P$.

A trait with an H^2 of 1.0 means that all of the variance in the phenotype is due to allelic variation of the genes. On the other hand, an H^2 of 0.0 means that all of the phenotypic variance is due to the environment. In a rough sense, H^2 can be understood to represent the extent to which a trait is genetically determined versus environmentally determined ("nature vs nuture"). Fisher partitioned the phenotypic variance V_P into additive V_A and dominant V_D components. The narrow-sense heritability h^2 is defined as $h^2 = V_A / V_P$.

The narrow-sense heritability allows one to predict how strongly a trait will respond to artificial selection, important in agriculture.

Test Yourself

1. Give an example of a trait not controlled at all by the environment.

2. In 1909, geneticist Herman Nilsson-Ehle showed that the color of wheat kernels is controlled by 3 genes, each with two incompletely dominant alleles. Predict the distribution of color resulting from a cross of $A'AB'BC'C \times A'AB'BC'C$, where every primed allele contributes one unit of redness and every unprimed allele contributes 0 units of redness (white).

3. Is the trait from question 2 continuous or discrete?

4. From question 2, what would be the probability of an individual from the cross having 4 units of red or more?

5. The wheat color trait is an example of genetic additivity. Explain what this means.

6. What is the form of the distribution for most quantitative traits?

7. What type of polygenic traits show a discrete presence or absence phenotype?

8. Give an example of a threshold trait.

9. How many genotypes are there for 5 genes, each with 2 alleles?

10. How many genotypes are there for 50 genes, each with 2 alleles?

11. What two factors result in a quantitative trait being continuous?

12. How can the number of genes underlying a discrete trait be ascertained?

13. For a cross of two fully heterozygous individuals, $1/256$ of the offspring have the trait associated with one of the homozygotes. How many genes control this trait?

14. What is the norm of reaction?

15. Calculate the mean for X_i = 10, 18, 14, 23, 25, 26, 30.

16. What is the variance and standard deviation for the data in question 15?

17. X_i = 10, 18, 14, 23, 25, 26, 30, and Y_i = 12, 22, 15, 24, 27, 30, 31. Calculate the covariance.

18. From the data in question 17, calculate the correlation coefficient.

19. Why not calculate the variance simply as the average of the deviations between each measurement and the mean, rather than the average of the squared deviations.

20. In a normal distribution, what does the peak of the curve correspond to?

21. What proportion of a normally distributed population lies between $\overline{X}$ +/− a single standard deviation? Two standard deviations?

22. For a population V_P = 6.2, V_G = 5.8, V_E = .4. What is H^2?

23. For a population V_P = 6.2, V_G = 5.8, V_E = .4, V_A = 3.8, V_D = 2.0. What is h^2?

24. Assume that for a certain species, the variance of the length of the forelimb is 504 (units not important). When all individuals are grown in the same environment, the variance reduces to 400. When many clones are grown in a variety of environments, the variance becomes 104. Assume the genetic/environmental covariance is 0. What is the broad-sense heritability?

25. From question 24, what would be V_E?

26. What exactly does V_A measure?

27. For a population, H^2 is measured as .40, and h^2 as .60. Is this possible?

28. What does the equation $V_P = V_G + V_E$ assume with respect to the interaction between genes and environment?

29. What is artificial selection?

30. How is artificial selection valuable for agriculture?

31. There are two cow breeds, one with h^2 = .30 for milk production, the other with h^2 = .60. Which population will be preferred by the dairy farmer?

32. Is heritability a fixed characteristic of a trait?

33. h^2 for bristle number in a population of *Drosophila* flies is .40. The mean number is 14. If a male and a female with means of 20 and 10 bristles are mated and a large number of offspring are produced, what will be the expected mean number of bristles among the offspring?

Check Yourself

1. Many simple Mendelian traits showing discrete variation are affected very little if at all by the environment. For example, pea plants carrying the dominant allele for flower color will be purple regardless of the environment in which they were raised. An example of a trait controlled by multiple genes but unaffected by the environment is given in question 2. **(Traits not affected by the environment)**

2. 1/64 0 units: 6/64 1 unit: 15/64 2 units: 20/64 3 units: 15/64 4 units: 6/64 5 units: 1/64 6 units. **(Simple, multiple gene traits)**

3. Discrete, because there are a countable number of trait categories. **(Simple, multiple gene traits)**

4. $15/64 + 6/64 + 1/64 = .34$. **(Simple, multiple gene traits)**

5. It means that the color of a kernel is simply the sum of the number of primed alleles. **(Simple, multiple gene traits)**

6. Normal curve (bell-shaped curve). **(Normal distribution)**

7. Threshold traits. **(Quantitative traits)**

8. Alcoholism and cleft lip are two examples. **(Quantitative traits)**

9. For a gene with 2 alleles, there are 3 genotypes. Therefore, for 5 such genes there are $3^5 = 243$ possible genotypes. **(The number of genotypes)**

10. $3^{50} = 7.2 \times 10^{23}$ possible genotypes. This illustrates that for a moderate number of genes, there will be an enormous number of possible genotypes. **(The number of genotypes)**

11. A continuous trait is one with an infinite number of phenotypic values. Two factors contribute to the number of values being uncountable. First, the enormous number of possible genotypes when there are a large number of genes controlling a trait, and second, the environment can produce a wide range of variation for a single genotype. **(Quantitative traits)**

12. If two heterozygotes are crossed, the fraction of offspring with one of the homozygous phenotypes provides a clue as to the number of genes involved. **(Simple, multiple gene traits)**

13. 4 genes. **(Simple, multiple gene traits)**

14. A graph that shows numerically how a fixed genotype responds to environmental variation. **(Quantitative traits)**

15. $\overline{X} = 20.9$. **(Basic statistics)**

16. $S^2 = 50.81$, $s = 7.1$. **(Basic statistics)**

17. $Cov_{XY} = 50.8$. **(Basic statistics)**

18. $r = .99$. The two variables are very strongly correlated. **(Basic statistics)**

19. Because if the deviations were not squared they would tend to sum to 0: $\Sigma(X_i - \overline{X}) = 0$. (**Basic statistics**)

20. The mean. (**Normal distribution**)

21. 66%, 95%. (**Normal distribution**)

22. $H^2 = 5.8/6.2 = .93$. (**Heritability**)

23. $h^2 = 3.8/6.2 = .62$. (**Heritability**)

24. $H^2 = 400/504 = .79$. (**Heritability**)

25. $V_E = 104$. (**Heritability**)

26. The additive variance component of the genetic variance; that portion of the genetic variance due to additive gene action rather than dominance and gene interaction. (**Heritability**)

27. No. The additive genetic variance must be equal to or less than the overall genetic variance; hence, $H^2 \geq h^2$. (**Heritability**)

28. That there is no interaction; in other words, that covGE = 0. What this means is that the contribution from an allele to the phenotype is independent of the environment. This is unlikely to be true in many situations. (**Heritability**)

29. Artificial selection is a process whereby certain animals or plants possessing desired characteristics are chosen to mate and produce the next generation. If the characteristic has a nonzero heritability, the trait will evolve in the desired direction. (**Artificial selection**)

30. Because it allows farmers to produce plants and animals giving greater yields. (**Artificial selection**)

31. The one with the higher h2, because this population will show a greater response to artificial selection. (**Artificial selection**)

32. No. Heritability depends on the population within which it is measured. A different population might give a different value for heritability if it possessed different alleles and/or was raised in a different environment. (**Heritability**)

33. $(15 - 14) (.40) + 14 = 14.4$. (**Heritability**)

Grade Yourself

Circle the numbers of the questions you missed, then fill in the total incorrect for each topic. If you answered more than three questions incorrectly, you need to focus on that topic. (If a topic has less than three questions and you had at least one wrong, we suggest you study that topic also. Read your textbook, a review book, or ask your teacher for help.)

Subject: Quantitative Genetics

Topic	Question Numbers	Number Incorrect
Traits not affected by the environment	1	
Simple, multiple gene traits	2, 3, 4, 5, 12, 13	
Normal distribution	6, 20, 21	
Quantitative traits	7, 8, 11, 14	
The number of genotypes	9, 10	
Basic statistics	15, 16, 17, 18, 19	
Heritability	22, 23, 24, 25, 26, 27, 28, 32, 33	
Artificial selection	29, 30, 31	

Structure of DNA

9

Brief Yourself

Elucidation of DNA as the Genetic Material

Early in this century, both nucleic acids (DNA, RNA) and proteins were considered as candidates for the genetic material. However, three separate sets of experiments ultimately proved that DNA is the molecule of heredity. These experiments were:

- 1928—transformation demonstrated in bacteria by Frederick Griffith,

- 1944—experiments by Oswald Avery, Colin MacLeod, and Maclyn McCarty prove that the transformation agent is DNA, and

- 1952—experiments by Alfred Hershey and Martha Chase prove that the genetic material in bacteriophages is DNA.

Structure of DNA

In 1953, the structure of DNA (deoxyribonucleic acid) was elucidated by James Watson and Francis Crick. Using chemical data from Erwin Chargaff and X-ray diffraction data provided by Maurice Wilkins, Rosalind Franklin, and others, Watson and Crick correctly deduced that DNA was a double-stranded helix. There are three levels to this model of DNA's structure. First, DNA is a polymer of molecules called nucleotides, which are distinguished from each other by their nitrogenous bases. There are four bases in DNA: adenine (A), thymine (T), guanine (G), and cytosine (C). Second, DNA is comprised of two strands of nucleotides, the two strands being held together by hydrogen bonds between bases on opposite strands. The base-pairing rules are the following: A always bonds to T, C always bonds to G. A further aspect of the second level of structure is that the two strands run in opposite directions, a feature referred to as antiparallel structure. Third, the two strands are twisted about each other to form a helix. DNA forms a double-helix.

Test Yourself

1. What is the first level of structure of DNA?

2. A nucleotide lacking phosphate groups is called a _____.

3. Is this a purine or pyrimidine?

4. What are the purine bases?

5. What are the pyrimidine bases?

6. Give the full names of the purine and pyrimidine deoxyribonucleoside triphosphates.

7. Give the full names of the ribonucleoside triphosphates (the RNA nucleotides).

8. What is the duplication of DNA called?

9. Draw the structure of a deoxyribonucleotide. Label the base, deoxyribose, and phosphate groups.

10. Draw deoxyribose and number the carbon atoms.

11. Draw the structure of ribose. What is the difference between ribose and deoxyribose?

12. Why are the carbon positions of ribose/deoxyribose labelled with primed numbers?

13. What is the correct flow of information?

    a. DNA→ protein→ RNA

    b. DNA→ RNA→ protein

    c. RNA→ protein→ DNA

    d. RNA→ DNA→ protein

14. Who first discovered DNA?

15. When heat-killed IIIS + living IIR *Diplococcus* bacteria are injected into mice, what happens?

16. Explain the outcome of the experiment in question 15.

17. What is the basis of transformation discovered by Griffith in 1928 as shown by the experiments of Avery, MacLeod, and McCarty in 1944?

18. How did the transformation experiments of Avery and colleagues differ from those of Griffith?

19. How could it be demonstrated that there was a transfer of genetic material from type III bacteria to type II bacteria, rather than a restoration of viability to type III by type II?

20. In 1952, Alfred Chase and Martha Chase performed an experiment where they showed that bacteriophages (viruses that parasitize bacteria) use DNA as their genetic material. They used phages labelled with ^{32}P and ^{35}S. What type of macromolecule would you expect to be labelled by each of these radioisotopes?

21. Who discovered that some viruses use RNA as their genetic material?

22. Where in the eucaryotic cell does DNA occur as opposed to RNA?

23. What are the three differences between DNA and RNA?

24. How does base-pairing work?

    a. purine-purine, pyrimidine-pyrimidine

    b. purine-pyrimidine

25. Why is the GC base-pair called the strong bond?

26. If [A] = 20%, what will [T] equal?

27. If [A] = 20%, what will [G] equal?

28. What did Erwin Chargaff's data show?

29. What was the significance of the X-ray diffraction data that Watson and Crick used?

30. What are the 3′ and 5′ ends of a single DNA strand?

31. Complete the following diagram by labelling the second strand with 3′ and 5′.

32. If the complete sequence of bases was known for one strand of DNA, the sequence of the other strand would be predicted. T or F?

33. The polymerization of two nucleotides results in the removal of a _____ molecule.

34. Why do the two strands of a DNA molecule have an antiparallel structure?

35. The core of a DNA molecule with the stacked base-pairs is:

    a. hydrophobic

    b. hydrophilic

36. Draw the structure of two polymerized deoxyribonucleotides.

37. What are the three types of DNA, and which is the most common form in nature?

38. How many DNA molecules make up a single chromosome?

39. A + T = C + G. T or F?

40. A + G = C + T. T or F?

41. What do renaturation kinetics with eucaryotic DNA demonstrate about eucaryotic DNA that is different from procaryotic DNA?

42. When two strands of DNA are separated by chemical or thermal means, this is known as _____.

 ## Check Yourself

1. The polymerized sequence of nucleotides forming a single strand of DNA. (**DNA general facts**)

2. Nucleoside. If we wish to emphasize the number of phosphate groups on a nucleotide, we adopt a nomenclature using the term nucleoside followed by the number of phosphate groups. For example, the building blocks of DNA are deoxyribonucleoside triphosphates. (**DNA terminology**)

3. This is the pyrimidine thymine. The pyrimidine bases are characterized by a single ring; the purine bases are characterized by two rings. (**Structure of DNA**)

4. Adenine and guanine. (**DNA general facts**)

5. Cytosine and thymine. (**DNA general facts**)

6. Deoxyadenosine triphosphate (the deoxyadenosine part refers to the nucleoside) (dATP), Deoxyguanosine triphosphate (dGTP), Deoxycytidine triphosphate (dCTP), Deoxythymidine triphosphate (dTTP). (**DNA terminology**)

7. Adenosine triphosphate (ATP), guanosine triphosphate (GTP), cytidine triphosphate (CTP), thymidine triphosphate (TTP). (**DNA terminology**)

8. DNA replication (see Chapter 10). **(DNA general facts)**

9.

phosphate
group   deoxyribose

(Structure of DNA)

10.

deoxyribose

(Structure of DNA)

11. Ribose has a hydroxyl group on the second carbon. Deoxyribose has a hydrogen in the same place; hence, the prefix "deoxy".

ribose

(Structure of DNA)

12. For the simple reason that the positions of the bases are numbered, and hence the numbers of the sugar must be distinguished with the prime (′) symbol. **(Structure of DNA)**

13. b. This flow of information is known as the Central Dogma in molecular biology. **(DNA general facts)**

14. Swiss medical student Johann Friedrich Miescher in 1868. **(Early work on DNA)**

15. The mice die. The type IIR bacteria were "transformed". **(Early work on DNA)**

16. Something from the dead IIIS cells transformed the living IIR cells to living IIIS cells. **(Early work on DNA)**

17. The transforming principle is DNA. **(Early work on DNA)**

18. Griffith never identified the transforming principle as did Avery and colleagues. **(Early work on DNA)**

19. Because purified DNA from type III cells was shown to be sufficient to transform type II cells. **(Early work on DNA)**

20. ^{32}P will be incorporated by DNA because phosphorus is a component of the phosphate groups, and ^{35}S will be incorporated by proteins because sulfur is a component of the side-chain of the amino acid cysteine. **(Early work on DNA)**

21. Heinz Fraenkel-Conrat and coworkers in 1957. **(Early work on DNA)**

22. In the eucaryotic cell, DNA is found in the nucleus; most RNA is found in the cytoplasm. **(DNA general facts)**

23. a. Deoxyribonucleotides lack oxygen on the 2′ position of the sugar

    b. DNA has T, whereas RNA has U (uracil), and

    c. DNA is double-stranded, RNA is single-stranded. **(Structure of DNA)**

24. b. **(DNA general facts)**

25. Because there are three H-bonds as opposed to the two bonds holding AT pairs together. **(Structure of DNA)**

26. 20%. **(Structure of DNA)**

27. If [A] = 20%, then [T] = 20%, and [A + T] = 40%, which means that [C + G] = 60%, and hence, [G] = 30%. **(Structure of DNA)**

28. That [A] = [T] and [C] = [G]. **(Early work on DNA)**

29. It showed that DNA has a helical structure and that the bases are stacked perpendicular to the axis of the molecule with a regular periodicity. **(Early work on DNA)**

30. The 3′ end is the free OH group on the 3′ deoxyribose carbon of the last nucleotide in a chain; the 5′ end is the free phosphate group(s) on the 5′ deoxyribose carbon of the first nucleotide in the chain. **(Structure of DNA)**

31.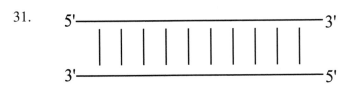

    (Structure of DNA)

32. T. **(DNA general facts)**

33. Water. All macromolecules in organisms are formed by dehydration reactions, which join monomers by the removal of water. **(DNA general facts)**

34. Because this is the only way for AT and CG base-pairs to form. **(DNA general facts)**

35. a. **(DNA general facts)**

36.

$O=P-O$
$O=P-O$

C — O — Base

H

$O=P-O$

C — O — Base

OH   H

(Structure of DNA)

37. A, B, and Z. B DNA is the most common form in cells. **(DNA general facts)**

38. Just one (but two strands of polymerized deoxyribonucleotides make up a single DNA molecule). **(DNA general facts)**

39. False. In fact, the CG content (relative to the AT content) is a basic descriptor of a stretch of DNA. **(DNA general facts)**

40. True. **(DNA general facts)**

41. Eucaryotic DNA has repeated sequences. Eucaryotic sequences fall into three categories:

    a. Unique or single-copy DNA sequences. 1–10 copies per genome,

    b. Moderately repetitive DNA sequences. 10–100,000 copies per genome, and

    c. Highly repetitive DNA sequences. More than 100,000 copies per genome. **(DNA general facts)**

42. Denaturation, which refers to the breaking of the H-bonds between bases on opposite DNA strands. **(DNA general facts)**

Grade Yourself

Circle the numbers of the questions you missed, then fill in the total incorrect for each topic. If you answered more than three questions incorrectly, you need to focus on that topic. (If a topic has less than three questions and you had at least one wrong, we suggest you study that topic also. Read your textbook, a review book, or ask your teacher for help.)

Subject: Structure of DNA

Topic	Question Numbers	Number Incorrect
DNA general facts	1, 4, 5, 8, 13, 22, 24, 32, 33, 34, 35, 37, 38, 39, 40, 41, 42	
DNA terminology	2, 6, 7	
Structure of DNA	3, 9, 10, 11, 12, 23, 25, 26, 27, 30, 31, 36	
Early work on DNA	14, 15, 16, 17, 18, 19, 20, 21, 28, 29	

DNA Replication

Brief Yourself

DNA Replication

Prior to 1958, the general nature of how a single DNA molecule replicates to form two new DNA molecules was unkown. Three mechanisms were considered possibilities: 1) conservative replication, whereby one of the daughter DNAs represents the original DNA while the other is entirely new DNA; or, 2) semiconservative replication, whereby both daughter DNA molecules are hybrids of old and new DNA, with one strand of a single daughter DNA being new DNA and the other strand representing original DNA; or, 3) dispersive replication, whereby both daughter DNAs are random mixtures of old and new DNA. The experiments of Matthew Meselson and Franklin Stahl in 1958 showed that the actual mechanism was semiconservative. The general mechanics of the process are fairly simple, involving base complementarity: the two strands of a DNA molecule are separated by breaking the H-bonds, and then the enzyme DNA polymerase brings in and polymerizes nucleoside triphosphates that are complementary to the two original strands. Although, the general mechanics of DNA replication are simple, the specific details are complex, involving a whole series of polypeptides.

Test Yourself

1. What phenomenon discovered by Watson and Crick in 1953 forms the basis for DNA replication?

2. Replication in eucaryotes occurs during the _____ phase of the cell cycle.

3. On the following diagram of a DNA strand, what will be the sequence of the other strand after replication?

   5'ATCCGATTA3'

4. From the diagram in question 3, in what direction will the new strand be synthesized?

5. All nucleic acids are synthesized in the $5' \rightarrow 3'$ direction. T or F?

6. How fast does the synthesis of new DNA occur in procaryotes?

7. How fast does the synthesis of new DNA occur in eucaryotes?

8. What are the three possible general mechanisms of DNA replication?

9. After fertilization, cell division produces an adult human with trillions of cells. Will all the cells contain absolutely identical DNA?

10. If cells are grown in ^{15}N for a long time, and then in ^{14}N for two generations, make predictions for the four daughter molecules of DNA produced (through those two generations) from each original DNA molecule in terms of the number of hybrid, heavy or light DNAs.

11. What is the bubble-like structure on the following replicating DNA molecule called? What are the structures indicated by arrows?

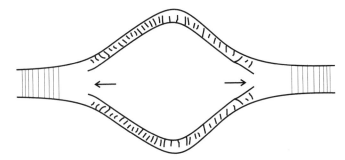

12. Provide an interpretation of the theta (Θ) structures found by John Cairns in 1963.

13. In procaryotes, how many origins of replication are there for the single chromosome?

14. In eucaryotes, there is only a single origin of replication per chromosome. T or F?

15. Replication is unidirectional or bidirectional?

16. The origin of replication in procaryotes is random and can occur anywhere along the chromosome. T or F?

17. What are the three activities of DNA polymerase I in *E. coli*?

18. What are the requirements for all DNA polymerases?

19. How many different DNA polymerases are known from *E. coli*?

20. What are the activities of DNA polymerase III from *E. coli*?

21. What is the specific function of DNA polymerase I in *E. coli*?

22. What is the specific function of DNA polymerase III in *E. coli*?

23. The following diagram shows a replication fork proceeding as shown by the arrow. Which strand will experience discontinuous replication?

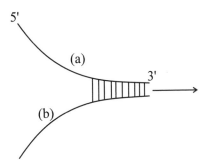

24. Which of the strands in question 23 will experience the most $5' \rightarrow 3'$ exonuclease activity of DNA polymerase I of *E. coli*?

25. From question 23, what are strands a and b called?

26. What are Okazaki fragments?

27. DNA ligase seals nicks due to discontinuous replication. Explain what a nick is.

28. What provides the energy for DNA replication?

29. What type of molecule primes the initial synthesis of new DNA?

30. How are RNA primers removed after DNA synthesis in *E. coli*?

31. The enzymes responsible for the unwinding of DNA that is necessary before replication can occur are known as _____.

32. The _____ proteins keep newly unwound DNA in the single-stranded configuration in *E. coli*.

33. The binding of SSB proteins is cooperative. What does this mean?

34. There are three DNA polymerases known from *E. coli*. Among eucaryotes, five DNA polymerases are known from mammals: α, β, γ, δ and ε. What are their functions?

35. Link each enzyme with its activity in *E. coli*.

    DNA gyrase   3′→ 5′ proofreading

    primase   removal of primers

    DNA polymerase I   synthesis of primers

    DNA polymerase III   negative supercoiling

36. Naturally occurring DNA molecules are negatively supercoiled. T or F?

37. What is the function of the DnaA protein in *E. coli*?

38. What is the primosome? What is the replisome?

39. What is a major problem that the linear chromosomes of eucaryotes face that the circular chromosomes of procaryotes do not during replication?

40. Which cells in eucaryotes have telomerase?

 # Check Yourself

1. Base complementarity. Once the two strands of a DNA molecule are separated, complementary nucleotides are brought in and polymerized together. (**Replication general facts**)

2. S. (**Replication general facts**)

3. 3′TAGGCTAAT5′. (**Replication general facts**)

4. From right to left. DNA is always synthesized in the 5′→ 3′ direction. (**Replication general facts**)

5. T. (**Replication general facts**)

6. About 30,000 nucleotides per minute. (**Rate of replication**)

7. Several thousand nucleotides per minute. (**Rate of replication**)

8. Conservative, semiconservative, and dispersive. (**General mechanisms of replication**)

9. No, because mutations occurring during development will lead to differences among the DNA molecules in different cells. (**Replication general facts**)

10. Conservative—1 heavy, 3 light strands; semiconservative—2 hybrid, 2 light; dispersive—4 hybrid. **(General mechanisms of replication)**

11. A replication bubble. The structures indicated by arrows, which form the ends of the bubble, are replication forks. **(Replication general facts)**

12. They represent procaryotic chromosomes caught in the act of replication. **(Replication general facts)**

13. Only one. **(Replication general facts)**

14. F. A eucaryotic chromosome is so large that many origins of replication are needed to replicate the entire chromosome within a reasonable amount of time. **(Replication general facts)**

15. Bidirectional, proceeding in both directions within a replication bubble. **(Replication general facts)**

16. F. There is a single fixed origin of replication. **(Replication general facts)**

17. The activities of *E. coli* DNA pol. I are:

    a. $5' \rightarrow 3'$ DNA synthesis,

    b. $5' \rightarrow 3'$ exonuclease activity, and

    c. $3' \rightarrow 5'$ exonuclease activity. **(DNA polymerases)**

18. The requirements for any DNA polymerase are:

    a. free deoxyribonucleoside triphosphates,

    b. a 3'-OH group on an already incorporated dNTP, and

    c. template strand. **(DNA polymerases)**

19. Three: DNA pol. I, II, and III. **(DNA polymerases)**

20. The activities of *E. coli* DNA pol. III are:

    a. $5' \rightarrow 3'$ DNA synthesis, and

    b. $3' \rightarrow 5'$ exonuclease activity (there is no $5' \rightarrow 3'$ exonuclease activity). **(DNA polymerases)**

21. To remove the RNA primers using the $5' \rightarrow 3'$ exonuclease activity and replace them with DNA. **(DNA polymerases)**

22. DNA synthesis and proofreading using the $3' \rightarrow 5'$ exonuclease activity. **(DNA polymerases)**

23. Strand a. **(Discontinuous replication)**

24. Strand a, because this is the strand that will have many RNA primers which will require removal by DNA pol. I. **(Discontinuous replication)**

25. The a strand is the lagging strand; the b strand is the leading strand. **(Discontinuous replication)**

26. An Okazaki fragment is a short stretch of DNA created by discontinuous DNA synthesis on the lagging strand. **(Discontinuous replication)**

27. A nick is simply a juncture between the 3′-OH group of one nucleotide and the 5′—phosphate group (or groups) of the next nucleotide. The nick is sealed when the two groups are covalently joined to form a phosphodiester bond. **(Discontinuous replication)**

28. The energy is provided by the hydrolysis of two phosphate groups off an incoming dNTP. **(Replication general facts)**

29. RNA. **(Replication general facts)**

30. By the $5′ \rightarrow 3′$ exonuclease activity of DNA pol. I. **(DNA polymerases)**

31. DNA helicases. **(Replication apparatus)**

32. Single-stranded binding (SSB) proteins. **(Replication apparatus)**

33. It means that the binding of one SSB protein facilitates the binding of others. **(Replication apparatus)**

34. δ leading strand replication

    α lagging strand replication

    γ organellar polymerase

    B, ε DNA repair. **(DNA polymerases)**

35. The proper connections are as follows:

    DNA gyrase————————negative supercoiling

    primase———————— synthesis of primers

    DNA polymerase I———— removal of primers

    DNA polymerase III———— $3′ \rightarrow 5′$ proofreading

    (Replication apparatus)

36. T. **(Replication general facts)**

37. It functions, along with a series of other proteins, to form the initiation complex and the replication forks. **(Replication apparatus)**

38. The primosome is a protein assembly containing DNA primase and DNA helicase. It is responsible for the unwinding and priming of the DNA. The replisome is a more inclusive protein assembly. It consists of the complete replication apparatus. **(Replication apparatus)**

39. After strand separation at a telomere (end of the chromosome), one tip of the lagging strand cannot be replicated because of the lack of a 3'-OH group on an already incorporated nucleotide after removal of the RNA primer. This leads to progressive DNA loss with every round of replication. **(Replication general facts)**

40. Germ-line cells and cancer cells. **(Replication general facts)**

Grade Yourself

Circle the numbers of the questions you missed, then fill in the total incorrect for each topic. If you answered more than three questions incorrectly, you need to focus on that topic. (If a topic has less than three questions and you had at least one wrong, we suggest you study that topic also. Read your textbook, a review book, or ask your teacher for help.)

Subject: DNA Replication

Topic	Question Numbers	Number Incorrect
Replication general facts	1, 2, 3, 4, 5, 9, 11, 12, 13, 14, 15, 16, 28, 29, 36, 39, 40	
Rate of replication	6, 7	
General mechanisms of replication	8, 10	
DNA polymerases	17, 18, 19, 20, 21, 22, 30, 34	
Discontinuous replication	23, 24, 25, 26, 27	
Replication apparatus	31, 32, 33, 35, 37, 38	

Transcription

11

Brief Yourself

Transcription

The flow of information from DNA to protein does not occur directly, but rather through the intermediate of messenger RNA (mRNA). This is known as the Central Dogma of molecular biology and is diagrammed as follows:

$$DNA \xrightarrow{\text{transcription}} mRNA \xrightarrow{\text{translation}} protein$$

In the process known as transcription, the deoxyribonucleotide code of the gene is converted to the ribonucleotide code of RNA. The code of the mRNA is then used to synthesize a protein by specifying the sequence of amino acids in a process known as translation. Transcription works by the same basic mechanism as DNA replication: base complementarity. The two DNA strands of a gene are locally unwound and then ribonucleotides complementary to one of the strands are brought in and polymerized by RNA polymerase. Like DNA synthesis, transcription operates in the $5' \rightarrow 3'$ direction. Important differences between transcription and DNA replication are that RNA contains the base uracil (U) instead of thymine (T) and bases contain ribose instead of deoxyribose.

Test Yourself

1. List three differences between RNA and DNA.

2. List three differences between transcription and DNA replication.

3. Give the sequence of RNA that will result from transcription of the following DNA.

   3′AATTCG5′

4. What are the four kinds of RNA and their functions?

5. On the following diagram, transcription proceeds as indicated by the arrow. Which strand will be transcribed (used as the template strand)?

   5′_____ a _____3′
   3′_____ b _____5′

6. From question 5, which strand is the sense strand and the antisense strand?

7. What is another term for the antisense strand?

8. Where on the DNA molecule does RNA polymerase initiate transcription?

9. How many RNA polymerases are known in procaryotes? In eucaryotes?

10. An mRNA transcript always corresponds to just one gene. T or F?

11. Very commonly, an mRNA corresponds to several genes in eucaryotes. T or F?

12. With regard to transcription, what do the terms "upstream" and "downstream" refer to?

13. What is the specific quaternary structure of RNA polymerase in *E. coli*?

14. What is the function of the σ factor from the RNA polymerase in *E. coli*?

15. What will happen if the core enzyme (no sigma factor) of *E. coli* RNA polymerase is introduced into a test-tube containing DNA and all the other requirements for transcription?

16. Transcription works in three sequential steps:
    1) _____, 2) _____, and
    3) _____.

17. Initiation of transcription involves three steps:
    a. binding of RNA polymerase to the promoter,
    b. localized unwinding of the DNA, and
    c. _____.

18. What change has to happen to the *E. coli* RNA polymerase holoenzyme before elongation can occur?

19. During initiation of transcription in *E. coli*, where does the transcription bubble form on the promoter?

20. The midpoints of the two conserved sequences of *E. coli* promoters occur at the –35 and –10 positions. What does this mean?

21. What is the significance of the fact that the –10 sequence is AT rich?

22. During elongation, how many ribonucleotides of the growing mRNA are H-bonded to the template strand at any one time?

23. What provides the energy for the synthesis of mRNA?

24. What are the two types of transcription terminators in *E coli*?

25. What is the requirement for rho-dependent termination?

26. Rho-independent termination in *E. coli* relies upon a palindromic sequence. A palindrome is a sequence of DNA in which the nucleotides of one strand are repeated in the reverse direction on the other strand. What can happen to a palindromic sequence after it is transcribed into RNA and that is especially critical in the context of termination of transcription?

27. For the following DNA sequence, complete the palindrome by adding 7 nucleotides to the 3′ end.

    5′ATTCGCC3′

28. List three major differences between procaryotic and eucaryotic transcription.

29. What are the three ways that a eucaryotic mRNA is posttranscriptionally processed?

30. How does termination of transcription work in eucaryotes?

31. List the function for the three eucaryotic RNA polymerases.

32. List two of the several conserved components of eucaryotic promoters.

33. What else beside RNA polymerase is required for initiation of transcription in eucaryotes?

34. What is an important difference between mRNA genes and tRNA and rRNA genes?

35. The enzyme responsible for adding the poly(A) tail to the 3′-end of the mRNA in eucaryotes is _____.

36. The genetic information in an mRNA is never altered after transcription. T or F?

37. In a eucaryotic interrupted gene, the coding sequences are called the _____, while the non-coding sequences are called the _____.

38. Most eucaryotic genes are interrupted. T or F?

39. The only completely conserved sequences in introns are two dinucleotide sequences that flank the 5′ and 3′ border of the sense strand. What are these sequences?

40. The 5′ consensus sequence for introns is $G_{100}T_{100}A_{68}A_{68}G_{84}T_{63}$ . . . . What does this mean?

Check Yourself

1. a. the ribonucleotides of RNA contain an OH group at the 2′ position, whereas DNA has only H at this position,

   b. RNA contains U instead of T, and

   c. RNA is single-stranded, DNA is double-stranded. **(General transcription)**

2. a. RNA is being made, not DNA (be sure you understand the differences between RNA and DNA),

   b. only one DNA strand is being copied, and

   c. there is no need on the part of RNA pol. for free 3′-OH's; thus, there is no need for primer sequences. **(Differences between DNA and RNA)**

3. 5′UUAAGC3′. **(General transcription)**

4. a. mRNA—transcript of protein-coding genes,

   b. rRNA—a component of ribosomes,

   c. tRNA—brings amino acids to the ribosome + mRNA, and

   d. snRNA (= small nuclear RNA)—component of spliceosome in eukaryotes. **(Types of RNA)**

5. The b strand, because RNA synthesis, like DNA synthesis, must occur in the 5′→ 3′ direction. **(General transcription)**

6. The a strand is the sense strand, the b strand the antisense strand. The sense strand is the one that is not used as the template for transcription. It is called the sense strand because it has the same sequence (except that there are T's instead of U's) as the mRNA. The antisense strand is the strand that is used during transcription. **(General transcription)**

7. Template strand. This is the strand used by the RNA polymerase as a guide during transcription. **(General transcription)**

8. On the promoter, a region that contains conserved sequences upstream from the gene. **(General transcription)**

9. There is one RNA polymerase in procaryotes, but three in eucaryotes. **(RNA polymerases)**

10. F. In procaryotes, often several genes are transcribed together. This gives rise to the term transcription unit to refer to a set of genes transcribed together. **(General transcription)**

11. F. Only one eucaryotic gene is transcribed at a time. **(General transcription)**

12. These terms refer to the direction of transcription. Upstream is in the direction opposite to RNA synthesis, downstream is in the same direction as RNA synthesis. **(General transcription)**

13. The quaternary structure of RNA polymerase in *E. coli* consists of 5 polypeptide chains representing 4 distinct proteins: $\alpha_2\beta\beta'\sigma$. **(RNA polymerases)**

14. The sigma factor functions to recognize and bind to the promoter upstream from the gene, thereby starting initiation. **(RNA polymerases)**

15. The core enzyme will begin transcription at random sites along the DNA, having lost its ability to bind specifically to promoters. **(RNA polymerases)**

16. Initiation, elongation, termination. **(General transcription)**

17. Formation of a small (several ribonucleotides) mRNA complementary to the start of the gene. **(Initiation of transcription)**

18. The sigma factor has to dissociate from the rest of the enzyme, otherwise the enzyme will not be able to travel down the gene. **(RNA polymerases)**

19. At the −10 conserved sequence. **(Initiation of transcription)**

20. This is part of a numbering system for individual nucleotides in reference to the start of the gene. Negative numbers correspond to the upstream region. **(Initiation of transcription)**

21. A region rich in A and T nucleotides dissociates easier because each AT base-pair is held together by only two H-bonds as opposed to the three bonds of GC pairs. **(Initiation of transcription)**

22. Only several, perhaps as few as three. **(Elongation of transcription)**

23. From the hydrolysis of two phosphate groups off each incoming ribonucleoside triphosphate. This is the same way that energy is provided for DNA synthesis. **(Elongation of transcription)**

24. Rho-dependent and rho-independent. **(Termination of transcription)**

25. The rho-protein. Rho-dependent terminators are sequences 50–90 bases long, rich in C's. The rho-protein binds to the mRNA transcript and moves along it in the $5' \rightarrow 3'$ direction until it encounters the RNA polymerase, which was slowed down by the termination sequence. At that point, termination is effected. **(Termination of transcription)**

26. It can base-pair on itself (intrastrand base-pairing). For example, the following DNA palindrome can form a base-paired RNA as shown.

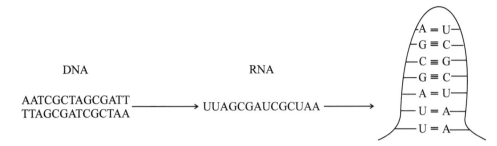

DNA

RNA

AATCGCTAGCGATT
TTAGCGATCGCTAA ⟶ UUAGCGAUCGCUAA ⟶

The significance of this in the context of transcription is that when the mRNA forms intrastrand H-bonds, it retards the movement of RNA polymerase, resulting in termination. (**Termination of transcription**)

27. 5′ATTCGCC + GCGAAT3′ forms a palindrome. The key is to add complementary nucleotides in reverse order to those already drawn. The student can easily verify that this is a palindrome by adding the second, complementary strand. (**Termination of transcription**)

28. a. In procaryotes, an mRNA can be multigenic, whereas in eucaryotes, it is almost always monogenic,

    b. in procaryotes, translation can begin as soon as transcription starts, but this cannot happen in eucaryotes where the mRNA must first diffuse out of the nucleus, and

    c. eucaryotic mRNAs are subject to posttranscriptional modifications. (**Differences between procaryotic and eucarotic transcription**)

29. a. 7-methyl guanosine caps are added to the 5′-ends of the mRNAs by a 5′ → 5′ phosphate linkage,

    b. poly(A) tails (20–200 A's) are added to the 3′-ends of the mRNAs, which are generated through cleavage and not termination sequences, and

    c. non-coding sequences called introns are cleaved out, and the exons spliced together. Eucaryotic genes are said to be split genes. (**Eucaryotic transcription**)

30. By cleavage, not by termination sequences. At some point, the 3′-end of the transcript is cleaved off, ending transcription. (**Eucaryotic transcription**)

31. a. RNA pol. I   rRNAs, excluding 5S rRNA

    b. RNA pol. II   mRNAs

    c. RNA pol. III   tRNAs, 5S rRNA, snRNAs (**RNA polymerases**)

32. There are several conserved motifs in eucaryotic promoters for protein-coding genes: octamer box, GC boxes, CAAT box, and TATA box. The octamer and GC boxes may be present or absent. (**Eucaryotic transcription**)

33. Transcription factors, proteins that interact with the promoter in a specific sequence to initiate transcription. (**Eucaryotic transcription**)

34. rRNA and tRNA genes are transcribed but not translated. **(Types of RNA)**

35. Poly(A) polymerase. **(Eucaryotic transcription)**

36. F. RNA editing is a process in eucaryotes that alters the mRNA in one of two ways: 1) by altering bases, or 2) by inserting or deleting uridine monophosphates. **(Eucaryotic transcription)**

37. Exons, introns. **(Eucaryotic transcription)**

38. T. Most but not all eucaryotic genes are interrupted. **(Eucaryotic transcription)**

39. 5′ exon—GT ... AG—exon 3′. **(Eucaryotic transcription)**

40. The letters represent the base that was the most common of the four bases at that position; the subscripted numbers represent the percentage of all examined sequences possessing that particular base. **(Eucaryotic transcription)**

Grade Yourself

Circle the numbers of the questions you missed, then fill in the total incorrect for each topic. If you answered more than three questions incorrectly, you need to focus on that topic. (If a topic has less than three questions and you had at least one wrong, we suggest you study that topic also. Read your textbook, a review book, or ask your teacher for help.

Subject: Transcription

Topic	Question Numbers	Number Incorrect
General transcription	1, 3, 5, 6, 7, 8, 10, 11, 12, 16	
Differences between DNA and RNA	2	
RNA polymerases	9, 13, 14, 15, 18, 31	
Initiation of transcription	17, 19, 20, 21	
Elongation of transcription	22, 23,	
Termination of transcription	24, 25, 26, 27	
Differences between procaryotic and eucarotic transcription	28	
Eucaryotic transcription	29, 30, 32, 33, 35, 36, 37, 38, 39, 40	
Types of RNA	4, 34	

Translation

12

Brief Yourself

Translation

Once the mRNA has been synthesized from the deoxy nucleotide code of a gene, its ribonucleotide code is used to direct the synthesis of a protein in the process of translation. Translation occurs on special organelles called ribosomes, which consist of a small and a large subunit. In addition to the mRNA and ribosome, translation requires a special kind of RNA called transfer RNA (tRNA), various proteins, and a source of energy. The tRNA brings the correct amino acids to the mRNA. In procaryotes, there is no nucleus separating the mRNA from the ribosomes, and therefore translation can occur simultaneously with transcription. In eucaryotes, the completed mRNA must diffuse out of the nucleus to dock with a ribosome, and therefore the processes of transcription and translation are decoupled.

The Genetic Code

Proteins are unbranched polymers of the twenty naturally occurring amino acids. In translation, amino acids are specified by sets of three nucleotides on the mRNA, called codons. The genetic code refers to the correspondence between all 64 possible codons and the amino acids (or other function) they specify. For example, the codon GUU (shown in ribonucleotides in the $5' \rightarrow 3'$ direction) codes for the amino acid valine. The code is uninterrupted and nonoverlapping, which means that there are no extra nucleotides between codons and that every nucleotide that is part of a codon is part of only that codon and no others. A remarkable feature of the genetic code is that, aside from a few exceptions in mitochondrial genomes, it is universal. This means that, regardless of the species, the same codon will always code for the same amino acid. For example, whether a bacterium or a human, GUU codes for valine.

Test Yourself

1. What are the three types of RNA involved in translation?

2. What is the role in translation of rRNA?

3. What is the role in translation of tRNA?

4. On the following diagram of a tRNA molecule, label parts a and b.

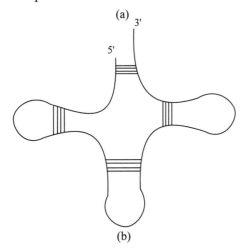

5. From question 4, give the function for parts a and b of the tRNA molecule.

6. Which end of the tRNA receives the amino acid?

7. List four of the unusual nucleosides that can be found in tRNAs and that result from posttranscriptional modification?

8. What is the exact nature of the linkage between the amino acid and the tRNA?

9. What provides the energy for charging the tRNA with an amino acid?

10. A charged tRNA is an _____.

11. The enzyme responsible for the charging of tRNAs is _____.

12. A tRNA is how large (in number of nucleotides)?

13. On the following diagram of a tRNA H-bonded to an mRNA transcript, label the 5′ and 3′ ends of the tRNA.

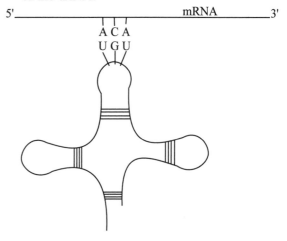

14. What allows there to be less than 64 tRNAs?

15. What will be the 5′→ 3′ sequence of the codons binding to the following anticodons (shown in the 3′→ 5′ direction): CCC, CGI, and AAU?

16. What sizes of rRNAs exist in the ribosomes of procaryotes?

17. What sizes of rRNAs exist in the ribosomes of eucaryotes?

18. The size of the small subunit in procaryotes is _____ and the large subunit is _____.

19. The size of the small subunit in eucaryotes is _____ and the large subunit is _____.

20. All translation in eucaryotes occurs on free ribosomes in the cytoplasm. T or F?

21. Explain the posttranscriptional modification of the eucaryotic rRNA gene.

22. In eucaryotes, there can be from hundreds to thousands of copies of the rRNA gene. Why?

23. In *E. coli*, how many repeats are there of the rRNA gene?

24. Ribosomes are constructed from two kinds of molecules, which are _____ and _____.

25. What part of the nucleus is the site of ribosome synthesis in eucaryotes?

26. Why is it impossible for codons to have evolved to be only two nucleotides long?

27. Explain what is meant by the statement that the genetic code is degenerate, nonoverlapping, and uninterrupted.

28. Give the amino acid coded for by each codon: UUU, UUA, AGU, GGU.

29. Which codon is the start codon and what amino acid does it code for?

30. Which codons are the stop codons? Which amino acids do they code for?

31. A remarkable feature of the genetic code is that it is nearly universal. What does this mean?

32. What are the exceptions to the universality of the genetic code?

33. What is a nonsense mutation?

34. What are the three binding sites for tRNAs on the ribosome?

35. The first AUG codon on the mRNA defines the start of translation in *E. coli*. T or F?

36. The first AUG codon on the mRNA defines the start of translation in eucaryotes. T or F?

37. What is the first amino acid to be incorporated into a protein in procaryotes?

38. Like transcription, translation can be broken into three sequential phases. What are they?

39. What are the steps of initiation in *E. coli*?

40. What are the steps of elongation in *E. coli*?

41. What is the fate of the initial methionine that begins a growing protein?

42. On the following diagram of a eucaryotic mRNA, give the exact sequence of the final protein that will be synthesized.

    5'UAUCUAAGGAUGAAAAAUCCACUUUAGUUUCCG3'

43. With reference to the protein synthesized by the mRNA in question 42, which end will correspond to the amino end (all proteins have a free amino end and carboxyl end)?

44. How long does the synthesis of a 600 amino acid protein take in *E. coli*?

45. What is the mechanism of termination in *E. coli*?

46. How does termination in eucaryotes differ from *E. coli*?

Check Yourself

1. mRNA—carries the code for synthesis of the protein, rRNA—a component of ribosomes, tRNA—brings amino acids to the translation machinery in reponse to the code of the mRNA. **(Types of RNA)**

2. rRNA is a constituent of ribosomes. In ribosomes, the rRNAs play several roles, including interaction with the mRNA through H-bonds and as enzymes. In procaryotes, the 5′-end of the 16S rRNA binds to the Shine-Dalgarno sequence on the mRNA. **(rRNA)**

3. tRNAs are said to be the adaptor molecules, which means that they mediate the specification of amino acids by the mRNA by bringing the correct amino acids to the mRNA during translation. **(tRNA)**

4. a. acceptor stem,

   b. anticodon. **(tRNA)**

5. a. acceptor stem—covalently binds to the amino acid,

   b. anticodon—binds to the codon of mRNA via H-bonding (base-pairing). **(tRNA)**

6. The 3′ end. **(tRNA)**

7. Pseudouridine, inosine (I), dihydrouridine, ribothymidine, methyl guanosine, dimethyl guanosine, and methyl inosine are some of the modified nucleosides found in tRNAs. **(tRNA)**

8. The carboxyl group of the amino acid is covalently linked to the 3′-OH of the nucleotide on the 3′-end of the tRNA. **(tRNA)**

9. The hydrolysis of 1 ATP. **(tRNA)**

10. Aminoacyl-tRNA. **(tRNA)**

11. Aminoacyl-tRNA synthetase. **(tRNA)**

12. 70–90 nucleotides. **(tRNA)**

13.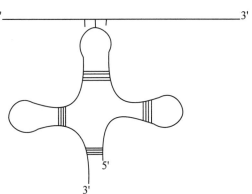

(tRNA)

14. The phenomenon of wobble, which is the ability of the 5′ position of the anticodon to bind to more than one nucleotide of the 3′ position of the codon. This allows a single anticodon (and hence tRNA) to correspond to several codons. **(tRNA)**

15. CCC — GGG; CGI — GCU, GCC, GCA; AAU — UUA, UUG. **(tRNA)**

16. 23S, 16S, 5S. **(rRNA)**

17. 28S, 18S, 5.8S, and 5S. Additionally, the eucaryotic organelles carry several genes for rRNAs. **(rRNA)**

18. 30S, 50S. **(Ribosomes)**

19. 40S, 60S. **(Ribosomes)**

20. F. Proteins that will be secreted by the cell are made on ribosomes bound to the membrane of the endoplasmic reticulum. **(Ribosomes)**

21. The eucaryotic rRNA gene is transcribed as a single unit; posttranscriptional modification then cleaves the rRNA precursor up to form the final rRNAs: **(rRNA)**

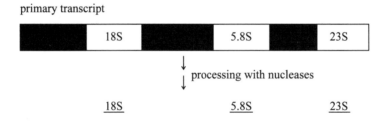

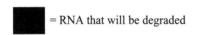 = RNA that will be degraded

22. Because this way many, many rRNAs can be made very quickly to meet the demand for ribosomes. **(rRNA)**

23. In *E. coli*, seven copies of the rRNA gene are distributed among three sites on the chromosome. **(rRNA)**

24. Proteins, RNA. **(Ribosomes)**

25. The nucleolus. This darkly staining region of the nucleus is the site where the rRNA genes are concentrated and is the site of ribosome assembly. **(Ribosomes)**

26. Because there would only be $4^2 = 16$ possible codons, which is fewer than the minimum of 20 required to code for all common amino acids. **(The genetic code)**

27. That the code is degenerate means that several codons can code for one amino acid. This has to be true because there are 64 codons, but only 20 amino acids. That the code is nonoverlapping means that any nucleotide that is part of a codon is part of no other codons. That the code is uninterrupted means that there are no nucleotides between codons; where one codon ends, the next codon starts without interruption. **(The genetic code)**

28. Phenylalanine, leucine, serine, and glycine. **(The genetic code)**

29. AUG, methionine. **(The genetic code)**

30. UAA, UAG, UGA. They code for no amino acid. Thus, 61 of the 64 codons code for amino acids. **(The genetic code)**

31. It means that for any given codon, that codon will specify the same amino acid regardless of the species. **(The genetic code)**

32. The mitochondrial genomes of certain species. **(The genetic code)**

33. A nonsense mutation is one that results in a stop codon in the middle of the coding sequence for a gene. Obviously, such a mutation will result in a nonfunctional protein since the latter half of the protein will be missing. **(The genetic code)**

34. From 5′ to 3′:

    a. E or exit tRNA binding site, which holds the exiting tRNA,

    b. P or peptidyl binding site, which holds the aminoacyl-tRNA with the growing protein, and

    c. A or aminoacyl binding site, which is the binding site for the incoming aminoacyl-tRNA. **(Ribosomes)**

35. F. The start of translation on the mRNA of *E. coli* corresponds to the first AUG after the Shine-Dalgarno sequence. **(Initiation of translation)**

36. T. **(Initiation of translation)**

37. Formylmethionine. This is methionine with a formyl group on the amino group. **(Initiation of translation)**

38. Initiation, elongation, and termination. **(General translation)**

39. Initiation in *E. coli* includes all events that precede the formation of the first peptide bond. Initiation involves the successive assembly of molecules—including the proteins known as initiation factors (IFs)—to form ever larger molecular complexes as follows:

    a. IF-2 + tRNA$_f$met,

    b. IF-3 + 30S subunit + mRNA,

    c. a + b + IF-1 + GTP, and

    d. c + 50S subunit. **(Initiation of translation)**

40. Elongation requires proteins called elongation factors (EFs) and works by a cyclical repetition of the following steps:

    a. an aminoacyl-tRNA with an anticodon complementary to the correct codon binds to the A site on the ribosome; this step requires EF-Tu charged with a GTP (EF-TuGTP),

    b. a peptide bond forms between the amino group of the amino acid on the A site and the carboxyl terminus of the growing protein on the tRNA of the P site. This reaction is catalyzed by peptidyl transferase. Peptide bond formation requires hydrolysis of the GTP carried by EF-Tu; EF-TuGTP is remade by EF-Ts,

    c. In the third step of elongation, the ribosome moves one codon in the $5' \rightarrow 3'$ direction, translocating the A site tRNA (now carrying the protein) to the P site, and the P site tRNA to the E site. Translocation requires GTP and EF-G. **(Elongation of translation)**

41. In both procaryotes and eucaryotes, the initial methionine is cleaved off. **(General translation)**

42. The secret to this question is to know exactly where translation will begin, which in eucaryotes is with the first AUG codon:

    5′UAUCUAAGGAUGAAAAAUCCACUUUAGUUUCCG3′

    start   stop

    The protein will be lysine—asparagine—proline—leucine. This example demonstrates the fact that for any mRNA, there will be 5′ and 3′ untranslated ends. **(General translation)**

43. The lysine end, because proteins are synthesized in the amino$\rightarrow$ carboxyl direction. Because of this, the amino end of a protein is taken to be the start of the protein. **(General translation)**

44. 30 sec. **(General translation)**

45. Termination in *E. coli* involves two protein release factors (RFs). Termination occurs when any of the three stop codons is encountered in the A site. RF-1 recognizes UAA and UAG; RF-2 recognizes UAA and UGA. The release factors cause the peptidyl transferase to cleave the growing polypetide chain from the P site tRNA by hydrolysis. **(Termination of translation)**

46. In eucaryotes, a single release factor (eRF) recognizes all three stop codons. **(Termination of translation)**

Grade Yourself

Circle the numbers of the questions you missed, then fill in the total incorrect for each topic. If you answered more than three questions incorrectly, you need to focus on that topic. (If a topic has less than three questions and you had at least one wrong, we suggest you study that topic also. Read your textbook, a review book, or ask your teacher for help.)

Subject: Translation

Topic	Question Numbers	Number Incorrect
Types of RNA	1	
rRNA	2, 16, 17, 21, 22, 23	
tRNA	3, 4, 5, 6, 7, 8, 9, 10, 11, 12, 13, 14, 15	
Ribosomes	18, 19, 20, 24, 25, 34	
The genetic code	26, 27, 28, 29, 30, 31, 32, 33	
Initiation of translation	35, 36, 37, 39	
General translation	38, 41, 42, 43, 44	
Elongation of translation	40	
Termination of translation	45, 46	

Proteins

13

Brief Yourself

Proteins

Except for rRNA and tRNA genes, genes code for proteins. Thus, genes accomplish their functions through the action of proteins. Proteins possess an amazing diversity of functions: many, perhaps most, are enzymes, whereas others function in immunity, transport, mechanical support, movement, communication, etc. Proteins are unbranched polymers of the 20 naturally occurring amino acids. All amino acids have two characteristic functional groups: a carboxyl group (—COOH) and an amino group (—NH_2). The 20 amino acids differ by their side groups, which range from nothing more than a hydrogen to organic molecules with ringed structures. Proteins possess four levels of structure: 1. primary, the exact sequence of amino acids; 2. secondary, due to hydrogen bonds between nearby amino acids; 3. tertiary, the three dimensional conformation of the protein, and; 4. quaternary, the presence of multiple polypeptides in a single protein. Not all proteins have a quaternary structure.

Test Yourself

1. On the following diagram of the amino acid alanine, label the carboxyl group, the amino group, and the side group.

$$
\begin{array}{c}
H \quad\quad H \quad\quad\quad O \\
\diagdown \quad\quad | \quad\quad\quad \diagup\!\!\diagdown \\
N-C-C \\
\diagup \quad\quad | \quad\quad\quad \diagdown \\
H \quad\quad CH_3 \quad\quad OH
\end{array}
$$

2. What are the four categories of amino acids based on the polarity of their side groups?

3. To which polarity category does the amino acid in question 1 belong to?

4. To which polarity category would an amino acid with a positively charged side group belong to? A negatively charged side group?

5. Is an amino acid with a hydrophobic side group charged or neutral at physiological pH?

6. In what direction with reference to the protein molecule does protein synthesis occur during translation?

7. On the following diagram of a dipeptide, which end would be considered the start of the protein?

(b) 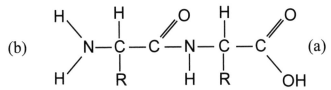 (a)

8. What is the range in protein size in nature in terms of the number of amino acids?

9. How many possible proteins will there be for 10 amino acids? For 100 amino acids?

10. What determines the chemical properties of an amino acid and hence how it determines the function of a protein?

11. What type of reaction polymerizes amino acids?

12. What are the two most common types of secondary structure?

13. The _____ structure of a protein will determine the _____ structure of a protein, which is the level that determines how a protein functions.

14. What happens when a protein is denatured?

15. What factors denature proteins?

16. What forces produce the tertiary structure of a protein?

17. With respect to the tertiary structure of proteins, what are the two shape classes?

18. Define each of the following terms:
    a. prosthetic group
    b. cofactor
    c. coenzyme
    d. apoenzyme
    e. holoenzyme

19. What is the prosthetic group for each of the following classes of conjugated proteins?
    a. metalloprotein
    b. lipoprotein
    c. glycoprotein
    d. flavoprotein
    e. phosphoprotein

20. What is the functional part of an enzyme?

21. Give a role for each of the following proteins:
    a. keratin
    b. hemoglobin
    c. antibody
    d. catalase
    e. actin

✔ Check Yourself

1.
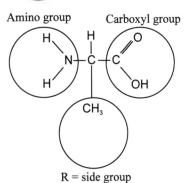
(Structure of amino acids)

2. a. hydrophobic

    b. hydrophilic

    c. basic

    d. acidic **(Structure of amino acids)**

3. This is alanine, which is hydrophobic. **(Structure of amino acids)**

4. An amino acid with a positively charged side group is basic because it has accepted a hydrogen ion; an amino acid with a negatively charged side group is acidic because it has lost a hydrogen ion. **(Structure of amino acids)**

5. A free amino acid will be neutral at physiological pH even though it carries a positive and negative charge on the amino and carboxyl groups, respectively. This type of schizophrenic ion is called a zwitterion. **(Structure of amino acids)**

6. From the amino terminus to the carboxyl terminus. **(General proteins)**

7. The end labelled b, which is the amino terminus. Because proteins are synthesized in the amino to carboxyl direction, the amino end is taken, by convention, to be the start of a protein. **(General proteins)**

8. From very small proteins like insulin (51 aa's) to proteins like fibroin (1000 aa's). **(General proteins)**

9. The number of possible proteins is 20^n, where n is the number of amino acids. For 10 amino acids, $20^{10} = 1.02 \times 10^{13}$, and for 100 amino acids, 20^{100} = a huge number. This underscores the fact that there can be a tremendous number of possible proteins for even moderate numbers of amino acids. This is why proteins can take on such a large number of roles. **(General proteins)**

10. The side group. **(Structure of amino acids)**

11. A dehydration reaction (the removal of a water molecule). **(General proteins)**

12. α helices and β sheets. **(Levels of protein structure)**

13. Primary, tertiary. **(Levels of protein structure)**

14. Its tertiary shape is changed, which alters or destroys the normal functioning of the protein. **(Levels of protein structure)**

15. High temperatures, extreme pHs are perhaps the most important factors that will denature proteins. **(Levels of protein structure)**

16. A series of mostly non-covalent bonds:

    a. ionic bonds

    b. H-bonds

    c. hydrophobic interactions

    d. covalent bonds (disulfide bridges between cysteine residues)

    e. Van der Waals interactions **(Levels of protein structure)**

17. Fibrous (e.g., keratin) and globular (e.g., hemoglobin). **(Levels of protein structure)**

18. a. prosthetic group—any non-peptide attached to a protein

    b. cofactor—any molecule associated with an enzyme and required for its functioning

    c. coenzyme—an organic cofactor (e.g., NAD+)

    d. apoenzyme—an enzyme without its cofactor

    e. holoenzyme—a complete enzyme, with every part necessary. **(General proteins)**

19. a. metalloprotein—metal

    b. lipoprotein—lipid

    c. glycoprotein—carbohydrate

    d. flavoprotein—flavin

    e. phosphoprotein—phosphate group(s). **(General proteins)**

20. The active site, a portion of the enzyme spanning several amino acids and that is the binding and catalytic site for the substrate(s). **(General proteins)**

21. a. keratin—waterproofing and mechanical support; forms hair, claws, stratum corneum, etc.

    b. hemoglobin—transport; carries oxygen

    c. antibody—immunity

    d. catalase—enzyme; breaks down hydrogen peroxide

    e. actin—movement; a component of muscle tissue **(General proteins)**

Grade Yourself

Circle the numbers of the questions you missed, then fill in the total incorrect for each topic. If you answered more than three questions incorrectly, you need to focus on that topic. (If a topic has less than three questions and you had at least one wrong, we suggest you study that topic also. Read your textbook, a review book, or ask your teacher for help.)

Subject: Proteins

Topic	Question Numbers	Number Incorrect
Structure of amino acids	1, 2, 3, 4, 5, 10	
General proteins	6, 7, 8, 9, 11, 18, 19, 20, 21	
Levels of protein structure	12, 13, 14, 15, 16, 17	

Control of Gene Expression

14

Brief Yourself

Control of Gene Expression

As seen in Chapter 12, the synthesis of proteins from the instructions in genes requires energy. Obviously then, it is to an organism's advantage to save energy by shutting off the genes whose products are not needed. An analogy would be shutting off the lights in your home when you go out. All species employ careful control of gene expression, which can occur on four levels: 1) transcription, 2) mRNA, 3) translation, 4) protein. In both procaryotes and eucaryotes, most gene regulation occurs at the transcriptional level; in other words, if a gene's product is not needed, it will not be transcribed. In procaryotes, the major theme of gene regulation centers around the operon, which is a set of genes that are transcribed as a unit, and hence regulated together. Operons are turned on or off by simple environmental signals. Thus, this represents temporal control of gene expression. In eucaryotes, control of gene expression is more complicated and involves an extra dimension in addition to temporal control: spatial gene regulation. Spatial control of gene expression comes in the form of cell differentiation as different cells express different sets of genes depending on their ultimate roles. Another way that the control of gene expression is more complicated in eucaryotes is that a eucaryotic promoter by itself is insufficient to cause RNA polymerase to initiate transcription. Various upstream sequences must be recognized by regulatory proteins, which bind to these sites, enabling RNA polymerase to initiate transcription.

Test Yourself

1. Some genes such as tRNA and rRNA genes must be turned on all the time in all cells. These are known as _____ genes.

2. Of the four levels of gene regulation, what level saves the most energy?

3. Bacterial genes for utilizing simple carbohydrates like glucose and lactose can be turned on very quickly in response to those molecules in the environment. These genes are said to be _____.

4. Genes that are turned off in response to environmental signals are _____.

5. In bacteria, inducible genes generally code for enzymes involved in _____ metabolism, whereas repressible genes generally code for enzymes involved in _____ metabolism.

6. Operons have three main components. What are they?

7. In an inducible operon like the lac operon, what happens when the effector molecule binds to the receptor?

8. The genes of an operon are all co-transcribed because they share only one _____.

9. On the following diagram of the *E. coli* lac operon, label the parts a-c.

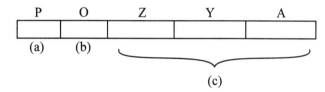

10. In the lac operon, the repressor gene codes for the repressor protein, which binds to the operator and shuts down transcription. This is an example of _____ control.

11. What is the enzyme and its function coded by each structural gene in the lac operon?

12. The lac repressor is an allosteric protein. What does this mean?

13. When the lac repressor binds to the operator, how does this block transcription?

14. What molecule is the effector molecule for the lac operon that binds to the repressor resulting in derepression?

15. What happens to the lac operon if both lactose and glucose are present?

16. The cAMP-CAP complex facilitates transcription of the lac operon. What kind of control is this?

17. The tryptophan (trp) operon is inducible/repressible.

18. What is the effector molecule controlling the trp operon?

19. The trp operon encodes how many structural genes?

20. How does the position of the trp repressor differ from the lac operon?

21. What general mechanism involved in transcription in *E. coli* is mimicked by the trp attenuation mechanism?

22. The trp attenuation mechanism relies on what type of sequence motif?

23. Explain how trp attenuation is defeated when tryptophan is rare.

24. What about the arabinose operon differs from the lac and trp operons?

25. Studies have shown that from the mRNA of an operon, different numbers of the proteins coded by the different genes are made. What kind of gene regulation is this?

26. What is a major difference in gene regulation between procaryotes and eucaryotes?

27. Spatial gene regulation occurs in all eucaryotes. T or F?

28. In what ways is gene expression more complicated in eucaryotes than in procaryotes?

29. What are trans- and cis-acting factors?

30. Gene regulation in eucaryotes works in what two dimensions?

31. What are two opposing cis-acting regulators of gene control in eucaryotes?

32. What distinguishes enhancers from promoters?

33. Before transcription can begin in eucaryotes, a preinitiation complex is required. This consists of RNA polymerase and _____.

34. Transcription factors have two critical domains. What are these?

35. In eucaryotes, what does the mechanism of alternate intron splicing accomplish?

36. Some eucaryotic mRNAs lack poly(A) tails. How do they differ from mRNAs with tails in terms of stability?

37. In procaryotes, genes can be induced by environmental factors. Give two examples of eucaryotic gene induction.

38. What are 4 types of DNA binding motifs in transcription factors?

39. Homeotic genes code for transcription factors that share a highly conserved helix-turn-helix DNA binding motif called the

    _____.

40. One mode of amplifying gene expression is exemplified by the rRNA gene in eucaryotes. Describe this mode.

41. What is the relationship between heterochromatin/euchromatin and gene expression in eucaryotes?

42. What disease results from the breakdown of the normal gene regulation of genes related to the cell cycle?

 # Check Yourself

1. Constitutive. (**General gene regulation**)

2. Regulation at the transcriptional level. The further that the process of translation is allowed to proceed, the more energy will be consumed. (**General gene regulation**)

3. Inducible. (**Procaryotic gene regulation**)

4. Repressible. (**Procaryotic gene regulation**)

5. Catabolic, anabolic. (**Procaryotic gene regulation**)

6. a. Promoter, the binding site for RNA polymerase,

   b. Operator, the binding site for the repressor. When the repressor is bound to the operator, RNA polymerase cannot transcribe the operon, and

   c. Structural genes, genes that code for enzymes. (**Procaryotic gene regulation**)

7. The repressor changes conformation and can no longer bind to the operator, thus allowing transcription. (**Procaryotic gene regulation**)

8. Promoter. (**Procaryotic gene regulation**)

9. a = promoter, b = operator, c = structural genes. (**Lac operon**)

10. Negative. (**Lac operon**)

11. Gene Z: β-galactosidase; cleaves lactose into glucose and galactose; gene Y: β-galactoside permease, brings lactose into cell, gene A: β-galactoside transacetylase, function unknown. (**Lac operon**)

12. Allosteric proteins change conformation when binding to another molecule. (**Lac operon**)

13. The operator is adjacent to the promoter and when the repressor is bound to the operator it blocks RNA polymerase's access to the promoter. (**Lac operon**)

14. Allolactose. (**Lac operon**)

15. In this situation, glucose is preferentially utilized and the lac operon is shut down. This mechanism, known as catabolite repression, works as follows: the lac promoter has two binding sites, one for a complex of catabolite activator protein and cyclic AMP (CAP-cAMP) and the other for RNA polymerase. CAP-cAMP must be present on its site in the promoter for transcription to occur (assuming of course that repressor is not binding the operator, in which case no transcription can occur). Glucose drives down the levels of cAMP, thus shutting down the lac operon. (**Lac operon**)

16. Positive. Positive control exists whenever the presence of a molecule(s) is required to turn on an operon. On the other hand, negative control is when the presence of a molecule turns off an operon. (**Lac operon**)

17. Repressible. (**trp operon**)

18. Tryptophan. Its presence represses the trp operon. The reason is simple: if tryptophan is present in the environment there is no reason to transcribe the trp operon, which codes for enzymes that synthesize tryptophan. (**trp operon**)

19. Five. (**trp operon**)

20. It is far removed from the trp operon, whereas in the lac operon, the repressor gene is nearby. (**trp operon**)

21. Rho-independent termination of transcription. (**trp operon**)

22. An inverted repeat or palindrome. (**trp operon**)

23. When there is tryptophan in the environment, transcription of the leader and attenuator regions results in a secondary structure with a conformation that causes termination of transcription, much as it occurs in rho-independent termination for genes. On the other hand, when tryptophan is rare or absent, a secondary structure forms that prevents the formation of the termination hairpin. This works in the following way: there are two trp codons (UGG) in a row on the leader. When translation of the leader mRNA begins at the 5′ end, the absence of tryptophanyl-tRNAs causes the ribosome to pause, preventing formation of the termination hairpin. (**trp operon**)

24. The product of the regulator gene can exert both negative and positive control over the operon. (**ara operon**)

25. Translational regulation. (**General gene regulation**)

26. Spatial gene regulation in eucaryotes. (**Eucaryotic gene regulation**)

27. F. Spatial gene control can only occur in multicellular organisms. (**Eucaryotic gene regulation**)

28. a. RNA processing (5′ cap, 3′ poly(A) tail, intron removal and splicing),

    b. compartmentalization; the mRNA must leave the nucleus and enter the cytoplasm to dock on ribosomes,

    c. organellar genomes; the mitochondrion and chloroplast both contain their own genomes, and

    d. cis-acting sequences (enhancers) and trans-acting factors (transcription factors) are required to enable RNA polymerase to initiate transcription. (**Eucaryotic gene regulation**)

29. A cis-acting factor is a sequence affecting a gene's transcription that is on the same chromosome as the gene. A trans-acting factor is a protein from the cytoplasm affecting a gene's transcription. (**Eucaryotic gene regulation**)

30. Spatial and temporal. (**Eucaryotic gene regulation**)

31. Enhancers and silencers. A silencer is a sequence that lowers the rate of transcription. (**Eucaryotic gene regulation**)

32. a. enhancers can act over long distances from the gene

    b. their influence on gene expression is independent of orientation, and

    c. their influence on gene expression is independent of position—they can be upstream or downstream of the gene. (**Enhancers**)

33. Transcription factors. For the pre-initiation complex, a whole series of transcription factors called basal transcription factors are required. (**Transcription factors**)

34. A domain that binds to DNA and one that facilitates transcription. (**Transcription factors**)

35. Alternate splicing is the phenomenon of removing successive introns in various ways to generate different proteins from the same gene. An example is the calcitonin peptide in the thyroid gland; an alternate splicing of the same gene gives rise to the CGRP peptide in the hypothalamus. Another example is the regulation of genes involved in Drosophila sex determination. Alternate splicing is gene regulation at the mRNA level. (**Alternate splicing**)

36. mRNAs that lack poly(A) tails are much less stable than mRNAs with tails. (**Eucaryotic gene regulation**)

37. a. Heat-shock proteins—these are proteins translated from genes that are turned on by heat stress, and

    b. genes regulated by hormones—hormones are molecules that induce the expression of certain genes. For example, steroid hormones are small hormones derived from the molecule cholesterol. Because of their lipid nature, they can easily pass through cell membranes. Once inside a cell, they bind to special receptors. The receptor-hormone combination now forms a transcription factor which will bind to certain genes regulating their expression. (**Eucaryotic gene regulation**)

38. a. Zinc finger motif

    b. helix-turn-helix motif

    c. leucine zipper motif

    d. helix-loop-helix motif **(Transcription factors)**

39. Homeodomain. **(Transcription factors)**

40. The rRNA gene in eucaryotes is repeated hundreds to thousands of times throughout the genome. This ensures that many copies of the rRNAs are available all the time. **(Eucaryotic gene regulation)**

41. Heterochromatin corresponds to DNA so tightly condensed that no gene expression occurs because RNA polymerase and other proteins controlling gene expression cannot access the DNA. Euchromatin, on the other hand, is loosely coiled DNA corresponding to regions containing actively expressed genes. **(Eucaryotic gene regulation)**

42. Cancer. Cancer results when a cell begins runaway cell division, developing into a mass of undifferentiated cells called a tumor. Cancer results from mutations to the genes that regulate cell division, which fall into two classes: proto-oncogenes, which stimulate cell division; and tumor-suppressing genes, which suppress cell division. A mutation to a proto-oncogene can convert it to an overactive form, resulting in runaway cell division. A mutation to a tumor-suppressing gene can inactivate it, resulting in runaway cell division. **(Eucaryotic gene regulation)**

Grade Yourself

Circle the numbers of the questions you missed, then fill in the total incorrect for each topic. If you answered more than three questions incorrectly, you need to focus on that topic. (If a topic has less than three questions and you had at least one wrong, we suggest you study that topic also. Read your textbook, a review book, or ask your teacher for help.)

Subject: *Control of Gene Expression*

Topic	Question Numbers	Number Incorrect
General gene regulation	1, 2, 25	
Procaryotic gene regulation	3, 4, 5, 6, 7, 8	
Lac operon	9, 10, 11, 12, 13, 14, 15, 16	
trp operon	17, 18, 19, 20, 21, 22, 23	
ara operon	24	
Eucaryotic gene regulation	26, 27, 28, 29, 30, 31, 36, 37, 40, 41, 42	
Enhancers	32	
Transcription factors	33, 34, 38, 39	
Alternate splicing	35	

Mutation

15

Brief Yourself

Mutation

A mutation is any permanent change to DNA. Mutations fall into two categories: point mutations, which are changes to one or a few nucleotides, and chromosomal mutations, which are large-scale changes to a chromosome that span multiple genes. Mutations are normally recognized by an altered phenotype, such as a change in the morphology of a eucaryote or the inability of a bacterium to synthesize a key nutrient. A mutation to a gene can be either dominant or recessive. A recessive mutation has to exist in the homozygous condition to be represented in the phenotype. The majority of mutations to genes are deleterious to the health of the bearer, producing genetic disease. However, occasional mutations are neutral or even beneficial. It is these mutations that form the raw material for evolution.

Causes of Mutation

Mutations can be either spontaneous or induced. Spontaneous mutations are mutations that occur without the influence of environmental agents. Induced mutations are caused by either chemical mutagens or radiation. A large number of potent mutagens have now been identified, a subset of which are carcinogens, chemicals that cause cancer by mutating genes involved in the regulation of the cell cycle.

Test Yourself

1. A mutation that so severely impacts a gene's function that the organism cannot survive is a _____ mutation.

2. A mutation occurring in a body cell is a _____ mutation, but a mutation occurring in a germ cell is a _____ mutation.

3. What is the critical difference between germinal and somatic mutations?

4. A fertilized zygote is homozygous for a mutation. Later in development, some groups of cells lack the phenotype associated with the mutation. Is this possible?

5. Very small changes to DNA are _____ mutations; large scale changes to chromosomes are _____ mutations.

6. What affects how many cells in an adult organism will carry a somatic mutation?

7. What are the three types of point mutations?

8. What kind of mutations do point insertions and deletions give rise to?

9. Point mutations are either spontaneous or

    _____.

10. The following diagram shows a short segment of DNA going through several rounds of replication, during which a mutation occurs. Circle the first point at which the mutation is permanent.

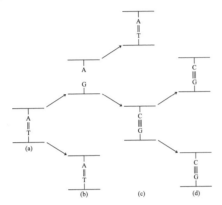

11. On the following diagram, circle the segment of DNA carrying the mutation.

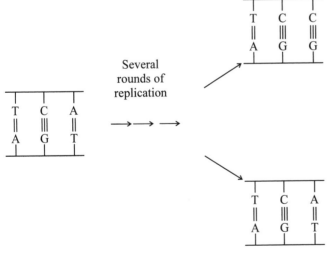

12. A chemical that induces mutations is a

    _____.

13. A cancer-causing chemical is a

    _____.

14. What are transition and transversion mutations?

15. List every possible transition and transversion.

16. Why are frameshift mutations so severe?

17. What general kind of mutations result from the following:

    a. proofreading errors by DNA polymerase

    b. tautomeric shifts

    c. slipped mispairing

    d. oxidative damage

18. The following are all examples of

    _____.

    a. alkylating agents

    b. base analogs

    c. acridine dyes

    d. deaminating agents

19. Acridine dyes are mutagens that induce what kind of point mutations?

20. Define each of the following categories of mutations:

    a. silent

    b. neutral

    c. missense

    d. nonsense

    e. suppressor

21. Which of the following mutations to a codon constitutes a missense mutation?

22. How does the frequency of missense mutations relate to the position of a nucleotide within a codon?

23. What is the molecular basis for the genetic disease sickle-cell anemia?

24. What is perhaps the most common cause of chromosomal deletions and insertions?

25. The following shows a change to the base cytosine that often leads to a mutation. What type of chemical change is this?

26. The radicals O_2^-, OH, and H_2O_2 result from normal metabolism and can cause _____ damage to DNA, resulting in spontaneous mutations.

27. Tautomers are isomers that differ in the positions of their atoms and bonds between atoms. All four bases of DNA can undergo tautomeric shifts spontaneously. For each base, give the base it pairs with when in its rarer tautomeric form (enol or imino).

28. What type of spontaneous substitutions do tautomeric shifts give rise to?

29. Under the Streisinger model of slipped mispairing for small spontaneous insertions and deletions, small loops can form on either strand of DNA during replication. If the loop occurs on the strand being synthesized, what kind of mutation results? If the loop occurs on the template strand, what kind of mutation results?

30. What type of mutations do the following phenomena give rise to?

    a. slipped mispairing

    b. acridine dyes

    c. transposable elements

    d. unequal crossing-over

31. What are expanding repeats and what causes them?

32. What are examples of genetic disease in humans caused by expanding repeats?

33. What kinds of radiation can cause mutations?

34. Why does visible light not cause mutations?

35. What is the difference in terms of mutation-causing potential between X–, gamma, and cosmic rays and UV radiation?

36. Can UV cause mutations deep in one's body?

37. What probably accounts for the different mutation rates among different species?

38. What causes pyrimidime dimers and what are the two ways that they cause mutations?

39. What are the five types of mutation repair in *E. coli*?

40. Describe the photoreactivation mechanism of repair for pyrimidine dimers.

41. What is the function of enzymes like superoxide dismutase and catalase?

42. What are two types of postreplication repair of DNA?

43. What causes the human disease xeroderma pigmentosum?

Check Yourself

1. Lethal. Mutations that impair an organism are deleterious. Lethal mutations are deleterious mutations that are so severe their bearers do not survive. **(General mutation)**

2. Somatic, germinal. **(General mutation)**

3. A germinal mutation can be passed on to offspring, a somatic mutation cannot. **(General mutation)**

4. Yes, because mutations are reversible. **(General mutation)**

5. Point, chromosome. **(Types of mutations)**

6. How early the mutation occurs in development. This is illustrated by the following diagram showing the course of development, where a mutation occurs early in development on the left, but later in development on the right.

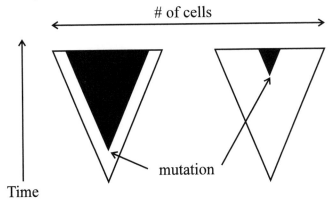

    (General mutation)

7. a. Substitution—the exchange of one nucleotide for another,

    b. insertion—the addition of one or several nucleotides, and

    c. deletion—the removal of one or several nucleotides. **(Types of mutations)**

8. Frameshift. A frameshift mutation is one where every codon downstream of the mutation is altered, giving rise to a completely altered protein. **(Types of mutations)**

9. Induced. **(Types of mutations)**

10. The mutation is made permanent at (c), because this is the point where a new, stable base pair (CG) is formed. Prior to this, an unstable base pair (AG) existed that would normally be subject to DNA repair. **(General mutation)**

11. The top segment, because it carries a base pair different from the original base pair. **(General mutation)**

12. Mutagen. **(Causes of mutations)**

13. Carcinogen. (**Causes of mutations**)

14. A transition is a change from a pyrimidine to another pyrimidine or from a purine to another purine. A transversion is a change from a pyrimidine to a purine, or vice-versa. (**Types of mutations**)

15. Transitions: A $\longleftrightarrow$ G, T $\longleftrightarrow$ C
    Transversions: A $\longleftrightarrow$ T, G $\longleftrightarrow$ C, T $\longleftrightarrow$ G, A $\longleftrightarrow$ C. (**Types of mutations**)

16. Because they alter the amino acid sequence of the protein dramatically, which will destroy the protein's original function. (**Types of mutations**)

17. These are all causes of spontaneous mutations. (**Causes of mutations**)

18. Mutagens. (**Causes of mutations**)

19. Frameshift mutations. (**Causes of mutations**)

20. a. silent—a substitution that conserves the amino acid,

    b. neutral—a mutation that does not affect the function of the protein,

    c. missense—a substitution that causes a codon to code for a new amino acid,

    d. nonsense—a mutation that forms a stop codon in the middle of a protein, and

    e. suppressor—a mutation that partially or completely cancels the effect of another mutation. (**Types of mutations**)

21. The second mutation. The first mutation is merely a conversion between two codons for leucine, whereas the second mutation converts a codon for leucine into one for proline. (**Causes of mutations**)

22. Substitutions at the first and second positions of codons are much more likely than substitutions at the third position to result in a new amino acid. This is simply a result of code degeneracy. (**Causes of mutations**)

23. Sickle-cell anemia results from a substitution to the sixth codon of the β-hemoglobin gene changing a valine to a glutamic acid. This demonstrates that a simple missense mutation can give rise to a genetic disease with a wide range of negative effects. (**Genetic disease resulting from mutations**)

24. Unequal crossing-over is perhaps the most common cause of large-scale insertions and deletions. (**Causes of mutations**)

25. Deamination (removal of the amino group). (**Causes of mutations**)

26. Oxidative. (**Causes of mutations**)

27. Imino form of C bonds to normal A
    enol form of G bonds to normal T
    enol form of T bonds to normal G
    imino form of A bonds to normal C (**Causes of mutations**)

28. Transitions. **(Causes of mutations)**

29. Insertion, deletion. **(Causes of mutations)**

30. Frameshift mutations. **(Causes of mutations)**

31. Expanding repeats are small sequences (such as CGG) that can duplicate themselves in tandem arrays. Slipped mispairing is the probable cause. **(Types of mutations)**

32. Fragile X syndrome, Kennedy's disease, and myotonic dystrophy (a type of muscular dystrophy). **(Genetic disease resulting from mutations)**

33. Very high energy radiation (X, gamma, and cosmic rays) and UV radiation. **(Causes of mutations)**

34. Visible light has insufficient energy. **(Causes of mutations)**

35. X–, gamma, and cosmic rays because of their very high energy can ionize molecules and thus have a greater potential for causing mutations than do UV rays, which cannot ionize molecules. **(Causes of mutations)**

36. No, because UV cannot penetrate very deeply into living tissue. **(Causes of mutations)**

37. Variation among species in the efficiency of the DNA repair machinery. **(DNA repair)**

38. Pyrimidine dimers are caused by UV radiation. They cause mutations by interfering with normal DNA replication or by errors that occur during the repair process. **(DNA repair)**

39. a. light-dependent repair (photoreactivation)

    b. excision repair

    c. mismatch repair

    d. postreplication repair

    e. error-prone repair **(DNA repair)**

40. The enzyme DNA photolyase detects and binds to pyrimidine dimers, absorbs visible light, and then cuts the bonds forming the dimer, restoring the original bases. **(DNA repair)**

41. To destroy radicals like O_2^-, OH, and H_2O_2 before they cause mutations. **(DNA repair)**

42. Mismatch repair, recombinational repair. **(DNA repair)**

43. An impairment of the excision repair system for UV caused pyrimidine dimers. **(DNA repair)**

Grade Yourself

Circle the numbers of the questions you missed, then fill in the total incorrect for each topic. If you answered more than three questions incorrectly, you need to focus on that topic. (If a topic has less than three questions and you had at least one wrong, we suggest you study that topic also. Read your textbook, a review book, or ask your teacher for help.)

Subject: Mutation

Topic	Question Numbers	Number Incorrect
General mutation	1, 2, 3, 4, 6, 10, 11	
Types of mutations	5, 7, 8, 9, 14, 15, 16, 20, 31	
Causes of mutation	12, 13, 17, 18, 19, 21, 22, 24, 25, 26, 27, 28, 29, 30, 33, 34, 35, 36	
Genetic disease resulting from mutations	23, 32	
DNA repair	37, 38, 39, 40, 41, 42, 43	

Recombination

16

Brief Yourself

Recombination

Recombination is the generation of novel combinations of genes. There are two kinds: intrachromosomal recombination or crossing-over, the (normally) reciprocal exchange of DNA between homologous chromosomes; and interchromosomal recombination, which is the mixing of chromosomes from different sources, such as occurs in diploid eucaryotes during meiosis. In procaryotes, crossing-over can occur whenever at least two homologous chromosomes coexist in the same cell, as might result from Hfr conjugation. Crossing-over in eucaryotes occurs during prophase I of meiosis when homologous chromosomes are paired up in the state called synapsis. Synapsis is maintained by the formation of the synaptinemal complex, a series of protein connections between chromosomes. Chiasmata (chiasma, s.) are the visible manifestations of crossing-over. One practical use of crossing-over is the construction of genetic maps. This is done under the simple principle that the further apart two genes are, the greater the probability that a cross-over event will occur between them.

Test Yourself

1. Can recombination occur intramolecularly? That is, within a single chromosome?

2. Predict the products of the following cross-over configuration for a pair of homologous chromosomes in meiosis.

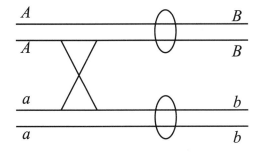

3. Predict the products of the following cross-over configuration for a pair of homologous chromosomes in meiosis.

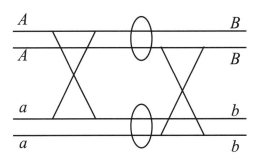

4. Predict the products of the following cross-over configuration for a pair of homologous chromosomes in meiosis.

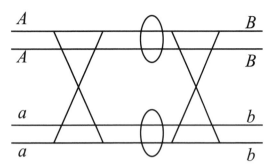

5. For the following two pairs of homologous chromosomes, what are the possible gametic genotypes given only interchromosomal recombination?

A	B		C	D
a	b		c	d

6. For the chromosomes of question 5, how many gametic genotypes will there be if there is both intra- and interchromosomal recombination?

7. Define interference.

8. The 1964 Robin Holliday model for crossing-over involves DNA strand breakage followed by strand reunion. What is the first step in the model?

9. In *E. coli*, what is the function of the RecBCD complex?

10. What is single-strand assimilation?

11. What is the role in *E. coli* of the RecA protein?

12. What is the role of DNA ligase in the Holliday model?

13. The Holliday model is the only mechanism by which crossing-over occurs. T or F?

14. Under the Holliday model, which of the following pair of homologs represent the true outcome of crossing-over?

or

15. During crossing-over, areas of hybrid DNA called _____ are formed.

16. Sometimes an allele is converted to another at a rate higher than the mutation rate. What explains this?

17. Crossing-over is always random, occurring anywhere along a chromosome. T or F?

18. During crossing-over, generally one or a few nucleotides are lost. T or F?

19. What type of mutations can result from errors in crossing-over?

20. On the following diagram of two homologous chromosomes, with which of the strands on the second chromosome will the indicated strand from the first chromosome undergo strand exchange?

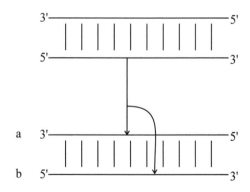

21. During meiosis, what happens to the position of chiasmata?

22. How does the Meselson-Radding model of crossing-over differ from the Holliday model?

23. The following diagram shows an F factor undergoing crossing-over with the circular chromosome of a bacterium. Draw the outcome.

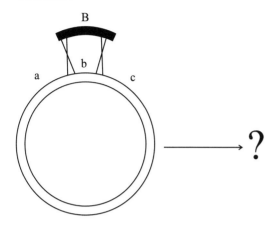

24. The following diagram shows a linear DNA fragment undergoing crossing-over with the circular chromosome in a bacterium. Draw the outcome.

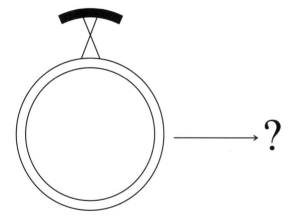

25. How many cross-overs would be required to integrate a linear DNA molecule into a circular DNA molecule?

26. How many cross-overs would be required to integrate two circular DNA molecules?

27. On the following diagram of a circular DNA molecule undergoing self-crossing-over, draw the outcome.

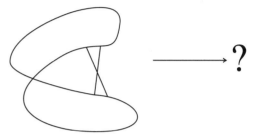

Check Yourself

1. Yes, intramolecular crossing-over can occur. All that is required are two homologous regions on the same chromosome. **(General crossing-over)**

2. From top to bottom: *AB, aB, Ab, ab*. **(General crossing-over)**

3. From top to bottom: *aB, Ab, Ab, aB*. **(General crossing-over)**

4. From top to bottom: *ab, AB, ab, AB*. **(General crossing-over)**

5. *ABCD, abcd, ABcd, abCD*. **(General crossing-over)**

6. 16. **(General crossing-over)**

7. Interference is the phenomenon whereby a cross-over at one site lowers the probability of cross-overs at other sites. **(Interference)**

8. A nick is made on a strand from each homologous chromosome by an endonuclease. **(Holliday model)**

9. The RecBCD complex contains both an endonuclease activity for introducing nicks to single DNA strands and a DNA helicase activity for unwinding the complementary strands adjacent to each nick. **(Holliday model)**

10. Single-strand assimilation is the process by which a single strand of DNA displaces its homolog on another chromosome. **(Holliday model)**

11. RecA promotes reciprocal strand assimilation between two homologous chromosomes. **(Holliday model)**

12. DNA ligase seals nicks that occur at two different times in the Holliday model: at the beginning and after the single strand bridge is resolved. **(Holliday model)**

13. F. Several models for crossing-over have been proposed. **(Alternate models of crossing-over)**

14. The right-hand pair, because crossing-over by the Holliday model will generate overlapping areas of hybrid DNA generated from both original chromosomes. **(Holliday model)**

15. Heteroduplex DNA. **(Holliday model)**

16. The phenomenon of gene conversion, which is the repair of mismatched bases in heteroduplex DNA. **(Holliday model)**

17. F. There are two kinds of recombination: generalized recombination, where cross-over can potentially occur almost anywhere along a chromosome; and site-specific recombination, which occurs only at specific places on a chromosome. An example of site-specific recombination is the λ phage chromosome, which is integrated into the *E. coli* chromosome at a specific site. **(General crossing-over)**

18. F. Crossing-over is usually exactly reciprocal, so that not even a single nucleotide is lost. **(General crossing-over)**

19. Insertions and deletions. **(General crossing-over)**

20. Strand b, because the $5' \rightarrow 3'$ directionality has to be maintained. (**General crossing-over**)

21. They move as the branch point migrates due to continued strand exchange. (**General crossing-over**)

22. Only one nick is made, followed by strand invasion of the nicked strand onto its homolog. (**Alternate models of crossing-over**)

23.

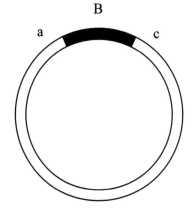

(**General crossing-over**)

24.

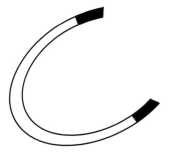

(**General crossing-over**)

25. Two. (**General crossing-over**)

26. One. (**General crossing-over**)

27. Two new circular DNA molecules will result. This happens during replicative transposition.

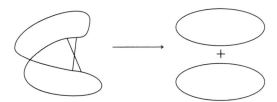

(**General crossing-over**)

Grade Yourself

Circle the numbers of the questions you missed, then fill in the total incorrect for each topic. If you answered more than three questions incorrectly, you need to focus on that topic. (If a topic has less than three questions and you had at least one wrong, we suggest you study that topic also. Read your textbook, a review book, or ask your teacher for help.)

Subject: Recombination

Topic	Question Numbers	Number Incorrect
General crossing-over	1, 2, 3, 4, 5, 6, 17, 18, 19, 20, 21, 23, 24, 25, 26, 27	
Interference	7	
Holliday model	8, 9, 10, 11, 12, 14, 15, 16	
Alternate models of crossing-over	13, 22	

Genetics of Viruses and Bacteria

17

 Brief Yourself

Viruses

Viruses are a class of microbes very distinct from bacteria. They are unique among organisms in lacking a cellular structure, and thus are considered neither procaryotes nor eucaryotes. Viruses have the simplest organization of any group of organisms, most being little more than a protein coat surrounding a single chromosome. Genetically, viruses are very simple, most having only a couple of dozen genes. As a result of their stripped-down structure, viruses are inert particles that do not metabolize. To reproduce, they must parasitize living cells and take control of those cells' replication and translation apparatuses. Viruses are classified by three criteria: 1) the type of nucleic acid they possess (DNA or RNA); 2) the strandedness of the chromosome; and 3) the presence or absence of a membranous envelope. While the majority of viruses have linear chromosomes, some have circular chromosomes. Because they parasitize cells, viruses are pathogens. Viruses that parasitize bacteria are called bacteriophages (or "phages"). A great deal is now known about the genetics of bacteriophages, especially the T phages and phage λ. Human viruses include such extremely pathogenic forms as the ebola virus and the HIV virus, which causes AIDS.

Phage λ

Normally, shortly after a phage injects its DNA into a bacterium, the bacterium lyses releasing new phage particles. Phage λ, however, is one of a number of phages that can also cause an infection without lysing the bacterium. This stable infection, called lysogeny, results when the λ chromosome is integrated into the *E. coli* chromosome. Once integrated, the λ chromosome—called the prophage—replicates along with the bacterial chromosome as the bacterium continues to grow and divide. Thus, once the λ chromosome is injected into an *E. coli* cell, it has two choices: to enter the lytic pathway or the lysogenic pathway. The choice is determined by the relative levels of regulatory proteins that control the expression of different genes.

Bacteria

Bacteria (kingdom Monera) are the simplest metabolizing cells. They consist of a cell wall, cell membrane, and the cytoplasm containing the single circular chromosome. In addition to the main chromosome, there can be one to several extrachromosomal DNAs called plasmids. Although bacteria reproduce asexually via mitosis, they are capable of transferring genetic material between cells. These parasexual processes include: 1) transformation, the uptake of environmental DNA by a bacterium; 2) conjugation, the transfer of DNA from one cell to another via specialized cell contacts; and 3) transduction, the transfer of DNA between cells via phages. A practical use of these processes is that they allow gene mapping.

Test Yourself

1. What immediately happens to the phage λ chromosome once it is injected into an *E. coli* cell? What feature of the λ chromosome allows this to happen?

2. What are temperate phages?

3. The λ chromosome can integrate itself anywhere on the bacterial chromosome. T or F?

4. What is the fate of the λ chromosome during a lytic vs. a lysogenic cycle?

5. During a lytic cycle, different sets of genes are sequentially turned on in three phases. Describe the three phases.

6. When integrated into the *E. coli* chromosome, the λ chromosome is called a _____.

7. What is the role of the N protein?

8. What is the role of the Q protein?

9. What are the two requirements for lysogeny?

10. How does the λ repressor work?

11. What are the functions of the two promoters P_{RE} and P_{RM} for the cI gene?

12. What enzyme coded by a late gene is responsible for bacterial lysis?

13. Describe the role for the regulatory proteins coded by each of the following genes:

    a. N

    b. Q

    c. cI

    d. cII

    e. cIII

    f. cro

14. What determines whether the phage enters the lytic or lysogenic cycle?

15. What is the function of the cro gene?

16. In what way is λ gene regulation similar to that in bacteria?

17. A phage capable only of lysis is _____.

18. Early studies on phage T4 suggested that the genetic map was circular, even though the chromosome was known to be linear. Explain this apparent contradiction.

19. The ΦΞ174 phage chromosome contains 5386 nucleotides and codes for genes that make proteins with a sum of 2300 amino acids. How is this possible?

20. The virus causing AIDS in humans, HIV, is a retrovirus. What does this mean?

21. What protection do bacteria employ against viral infection?

22. What are the three parasexual processes in bacteria?

23. Is the exchange of DNA during parasexual processes reciprocal?

24. Define auxotroph, prototroph.

25. Is transformation a passive process?

26. What bacterial cells can be transformed?

27. What is the significance of transformation to the history of molecular genetics?

28. Describe the events of transformation.

29. Can genes be mapped based on transformation?

30. In 1946, Joshua Lederberg and Edward Tatum mixed auxotrophic strains of *E. coli* (met⁻bio⁻thr⁺leu⁺ and met⁺bio⁺thr⁻leu⁻). Later, they found that prototrophic bacteria had developed. Explain.

31. What is required for a cell to be a donor in conjugation?

32. During conjugation, the F factor is transferred to the recipient cell by rolling circle replication. Explain rolling circle replication.

33. What is an episome?

34. What is sexduction?

35. How is sexduction useful in genetic studies?

36. What is Hfr conjugation?

37. Interrupted Hfr conjugation experiments can be used to map genes. The maps that result report distances in units of minutes. Explain.

38. From two Hfr gene mapping experiments, the following gene orders are obtained: ACDBE and EBDCA. Is this possible?

39. Using the following gene orders from three Hfr mapping experiments, deduce the order of the genes.

    EHGA

    BDFE

    FDBCAG

40. What are the two types of transduction?

41. How does gene mapping based on transduction work?

Check Yourself

1. The molecule circularizes because of single-stranded, complementary ends. (**λ phage genetics**)

2. Temperate phages are capable of lysogeny. (**General phage genetics**)

3. F. The λ chromosome inserts itself between the *E. coli* gal and bio loci. (**λ phage genetics**)

4. In the lytic cycle, the λ chromosome remains free in the cytoplasm; during lysogeny, the λ chromosome becomes stably incorporated into the bacterial chromosome. (**λ phage genetics**)

5. a. Immediate-early—transcription begins on both strands in the left and right directions at the P_L and P_R promoters using the host's RNA polymerase. The N and cro genes are transcribed during this phase,

   b. delayed-early—the N protein acts as an antiterminator to allow transcription to continue past the N and cro genes. This results in the transcription of the cII and cIII, as well as the O, P, and Q genes. The O and P gene products replicate the viral DNA, and

   c. late phase—the Q protein is another antiterminator which allows transcription of the late genes, including genes for the head and tail proteins and for lysis. (**λ phage genetics**)

6. Prophage. (**λ phage genetics**)

7. To act as an antiterminator and allow transcription to proceed past the N and cro genes, allowing transcription of the delayed-early genes. (**λ phage genetics**)

8. To act as an antiterminator and allow transcription to proceed past the t_{R3} termination site, allowing transcription of the late genes. (**λ phage genetics**)

9. The λ repressor, coded by the cI gene, and the Int protein, coded by the int gene. The Int protein is responsible for integration of the phage chromosome into the bacterial chromosome. (**λ phage genetics**)

10. The λ repressor, coded by the cI gene, is a dimer that binds to the operators O_L and O_R, blocking transcription at the promoters P_L and P_R. The dimers have the highest affinity for the O_{L1} and O_{R1} sites. The presence of dimers at these sites generates cooperative binding of repressor to the next pair of operator sites (O_{L2} and O_{R2}). However, this cooperativity does not extend to the third pair of sites (O_{L3} and O_{R3}). (**λ phage genetics**)

11. The P_{RE} promoter is responsible for the initial synthesis of repressor. The initial repressor dimers then bind to the O_{L1}, O_{L2}, O_{R1}, and O_{R2} sites, which actually stimulates transcription of more dimer from the P_{RM} promoter, ensuring the stability of lysogeny. Thus, the P_{RM} promoter is responsible for the maintenance of repression. (**λ phage genetics**)

12. Lysozyme. (**λ phage genetics**)

13. a. N—codes for the antiterminator that allows transcription of the delayed-early genes

    b. Q—codes for the antiterminator that allows transcription of the late genes

    c. cI—codes for the λ repressor protein that binds to the left and right promoters

    d. cII—promotes transcription from the P_{RE} promoter for the cI gene

    e. cIII—stabilizes cII protein

    f. cro—codes for a repressor protein that binds to the O_L and O_R sites. (**λ phage genetics**)

14. Whether the phage enters the lytic or lysogenic cycle is determined by a balance between two regulatory proteins: cro and λ repressor. The cro protein is perhaps the more pivotal. Synthesized soon after the λ chromosome enters the bacterium, the cro protein is a repressor that binds to the O_L and O_R sites. However, unlike the λ repressor, the cro protein has a stronger affinity for the O_{L3} and O_{R3} sites, where it prevents transcription of the cI gene from the P_{RM} promoter. Also, unlike the λ repressor, the cro protein is not a completely efficient repressor, allowing some transcription to occur leading to lysis. On the other hand, if there is a greater balance of λ repressor initially, a self-sustaining lysogenic phase will ensue. This is because the initial λ repressor will bind the O_{L1}, O_{L2}, O_{R1}, and O_{R2} sites, preventing transcription of the delayed-early and late genes, but facilitating lysogeny by stimulating transcription of the cI gene from the P_{RM} promoter. (**λ phage genetics**)

15. The cro protein is a repressor that binds to the O_L and O_R sites. However, unlike the λ repressor, the cro protein has a stronger affinity for the O_{L3} and O_{R3} sites, where it prevents transcription of the cI gene from the P_{RM} promoter. Also, unlike the λ repressor, the cro protein is not a completely efficient repressor, allowing some transcription to occur leading to lysis. (**λ phage genetics**)

16. The genes of phage λ are regulated together from shared promoters, just like the operon system of bacteria. (**λ phage genetics**)

17. Virulent. (**General phage genetics**)

18. This is explained by the fact that the T4 chromosome is terminally redundant (ends carry same genes) and circularly permuted. Basically, before packaging into the new phage heads, many of the linear phage chromosomes are concatenated together to form a giant linear molecule. The molecule is stuffed into a phage head, and then cleaved when the head is full. This leaves a few extra genes on the end, hence the redundant ends. Each chromosome will be cleaved at different points, generating the phenomenon of circular permutation. (**T4 phage genetics**)

19. Because genes in viruses can be overlapping, both on the same strand or different strands. (**General phage genetics**)

20. Retroviruses carry RNA genomes that are reverse-transcribed upon infection in a cell to DNA. The conversion is catalyzed by the enzyme reverse transcriptase. (**Retroviruses**)

21. Endonucleases. These are enzymes that cleave DNA at specific 4–8 nucleotide palindromes. They function to destroy incoming viral DNA. (**General bacterial genetics**)

22. a. Transformation, the uptake of environmental DNA by a bacterium,

    b. Conjugation, the transfer of DNA from one cell to another via specialized cell contacts, and

    c. Transduction, the transfer of DNA between cells via phages. **(General bacterial genetics)**

23. No, it is unidirectional. **(General bacterial genetics)**

24. An auxotroph is a mutant bacterium that lacks the ability to synthesize a nutrient and thus must be provided this nutrient. A prototroph can synthesize all the essential nutrients and hence can exist on a minimal medium. **(General bacterial genetics)**

25. No, it requires energy. The energy is provided by the hydrolysis of one of the two strands of a DNA molecule being taken up by a recipient cell. **(Transformation)**

26. Competent cells. Competent cells possess the competence factor, a small protein that induces the synthesis of proteins involved in the uptake of DNA. **(Transformation)**

27. Transformation was first discovered by Frederick Griffith in 1928. This led to the identification of the transforming principle as DNA, thereby identifying DNA as the genetic material. **(Transformation)**

28. a. DNA binds specific receptors on recipient cell,

    b. One strand is hydrolyzed, generating a single strand,

    c. DNA moves across cell membrane, and

    d. DNA recombines with homologous site on host chromosome. **(Transformation)**

29. Yes, by the frequency that pairs of genes are co-transformed. **(Transformation)**

30. Genetic recombination by conjugation. **(Conjugation)**

31. The bacterium must possess the F factor, a plasmid carrying the genes for cell-to-cell connections and F factor transfer. **(Conjugation)**

32. Rolling circle replication begins with a nick to a circular chromosome. The nicked strand is displaced from its complementary strand and begins migration into the recipient cell, resulting in rotation of the chromosome. As the nicked strand continues to pull away from the other strand, DNA synthesis replaces it. Eventually, a double-stranded chromosome exists in the donor cell and a single-stranded chromosome in the recipient cell. **(Conjugation)**

33. An episome is a plasmid that can exist freely in the cytoplasm or integrated in the main chromosome. **(Conjugation)**

34. Sexduction is when some chromosomal genes are transferred along with the F factor during Hfr conjugation. This results in the F′ factor. **(Conjugation)**

35. Sexduction can be used to map genes or, because partial diploidy is generated in the recipient cell, to study allelic interactions. **(Conjugation)**

36. Hfr conjugation results when the F factor integrates itself into the donor cell's chromosome before transfer to the recipient cell. **(Conjugation)**

37. Because the map distances are measured by the amount of time it takes each gene to appear in the recipient cell. **(Conjugation)**

38. Yes, because the F factor can insert itself anywhere in a chromosome and in both orientations. **(Conjugation)**

39. The map is circular:

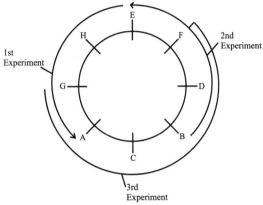

    (Conjugation)

40. a. Generalized transduction, the carrying by a phage of any bacterial gene(s), and

    b. Specialized transduction, the carrying only of specific portions of the bacterial genome. **(Transduction)**

41. By the frequency that genes are co-transduced. **(Transduction)**

Grade Yourself

Circle the numbers of the questions you missed, then fill in the total incorrect for each topic. If you answered more than three questions incorrectly, you need to focus on that topic. (If a topic has less than three questions and you had at least one wrong, we suggest you study that topic also. Read your textbook, a review book, or ask your teacher for help.)

Subject: Genetics of Viruses and Bacteria

Topic	Question Numbers	Number Incorrect
λ phage genetics	1, 3, 4, 5, 6, 7, 8, 9, 10, 11, 12, 13, 14, 15, 16	
General phage genetics	2, 17, 19	
T4 phage genetics	18	
Retroviruses	20	
General bacterial genetics	21, 22, 23, 24	
Transformation	25, 26, 27, 28, 29	
Conjugation	30, 31, 32, 33, 34, 35, 36, 37, 38, 39	
Transduction	40, 41	

Transposable Elements

18

Brief Yourself

Transposable Elements

As should be clear now, genomes are very dynamic. The processes of mutation and recombination, which have already been encountered in earlier chapters, result in dramatic changes to genomes through time. Additionally, there exist DNA sequences called transposable elements that are capable of "jumping" or transposing from one part of a genome to another. There are two modes of transposition: replicative, in which the transposon leaves a copy of itself at the original site; and conservative, in which the transposon leaves the original site and moves to a new site in the genome. A transposable element—or more simply, a transposon—can lead to a frameshift mutation if it lands in the middle of a gene. In fact, this is how transposons were originally discovered—by mutant phenotypes due to frameshift mutations induced by transposition. Transposons were originally discovered in maize by Barbara McClintock, but today procaryotic transposable elements are better understood. There are some eucaryotic transposons—called retrotransposons—whose movement is similar to the reproduction of retroviruses.

Test Yourself

1. Transposons can only move to a new site on the same chromosome. T or F?

2. What are the three types of transposons in bacteria?

3. What distinguishes composite transposons and Tn3 elements from IS elements?

4. What distinguishes composite transposons and Tn3 elements?

5. How were IS elements first detected in bacteria?

6. What are the two parts of an IS element?

7. How do the inverted repeats of IS elements differ from palindromes?

8. What characterizes the insertion site for IS elements after transposition has occurred?

9. How does target site duplication arise?

10. Of the following pairs of sequences, which could qualify as inverted repeats?

    a. 5'-AATTCGCGT-3' and 5'-AATTCGCGT-3'

    b. 5'-GAATCCGCA-3' and 5'-TGCGGATTC-3'

    c. 5'-AATTCGCGT-3' and 5'-AATTCGAAC-3'

    d. 5'-GAATCCGCA-3' and 5'-CCAAGATTC-3'

11. Of the following pairs of sequences, which could qualify as direct repeats?

    a. 5'-AATTCGCGT-3' and 5'- AATTCGCGT-3'

    b. 5'-GAATCCGCA-3' and 5'-TGCGGATTC-3'

    c. 5'-AATTCGCGT-3' and 5'- AATTCGAAC-3'

    d. 5'-GAATCCGCA-3' and 5'-CCAAGATTC-3'

12. Label the parts a-c on the following diagram of an IS.

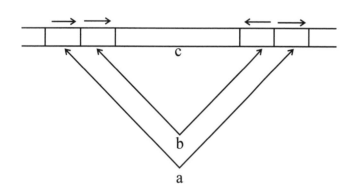

13. What are the parts of a composite transposon?

14. The flanking IS's of a composite transposon are always in the same orientation. T or F?

15. Composite transposons and Tn3 elements carry genes unrelated to transposition. What do they do?

16. What is the medical significance behind transposition in pathogenic bacteria?

17. What protein catalyzes procaryotic transposition?

18. How do the replicative and conservative modes of transposition differ?

19. What aspect of phage mu reproduction is similar to transposition?

20. Which were the first transposable elements discovered?

21. Why is the Ac transposon required for Ds activation in maize?

22. Barbara McClintock discovered transposons through her work with mutant phenotypes of corn kernels. Many kernels that she examined were spotted purple and yellow. Explain the production of this phenotype.

23. Transposons form insignificant portions of eucaryotic genomes. T or F?

24. How are the Ds and Ac maize transposons related?

25. What is a retrotransposon?

26. What are processed pseudogenes?

27. How can transposons affect chromosomal evolution?

28. Explain hybrid dysgenesis in *Drosophila*.

 Check Yourself

1. F. They can move between chromosomes as well. (**General transposons**)

2. a. insertion sequences (IS elements)

    b. composite transposons

    c. Tn3 elements (**Procaryotic transposons**)

3. Composite transposons and Tn3 elements, unlike IS elements, carry genes unrelated to transposition. **(Procaryotic transposons)**

4. Tn3 elements are simpler than composite transposons. They lack IS's at each end. Instead, Tn3 elements are characterized by inverted repeats framing a region carrying genes for transposition and (usually) genes unrelated to transposition. **(Procaryotic transposons)**

5. By the frequent reversion of lac⁻ mutants. Reversion in these mutants was due to IS elements moving out of the lac operon, thereby repairing the frameshift mutation caused by the presence of the extra DNA. **(Procaryotic transposons)**

6. Inverted terminal repeats and a region containing the genes for transposition. **(Procaryotic transposons)**

7. The inverted repeats of IS's are essentially palindromes but differ from them in being interrupted by the transposition genes. **(Procaryotic transposons)**

8. There is a short duplication (2 to 13 nucleotide pairs) on either side of the transposon. These are called direct repeats. **(Procaryotic transposons)**

9. By several steps:

   a. Staggered cleavage at the target site,

   b. Insertion of the IS, and

   c. Replication of the single-stranded DNA. **(Procaryotic transposons)**

10. b., because the second sequence in the pair is the reverse complement of the first. **(Procaryotic transposons)**

11. a., because these sequences are identical. **(Procaryotic transposons)**

12. a. Direct repeats at target site,

   b. Inverted repeats, and

   c. Genes for transposition. **(Procaryotic transposons)**

13. Composite transposons have three parts: two flanking IS's and a middle region containing genes for transposition and other functions. **(Procaryotic transposons)**

14. F. The flanking IS's can be in either orientation. For example, Tn has flanking IS's oriented in the same direction, but Tn5 has flanking IS's oriented in opposite directions. **(Procaryotic transposons)**

15. They are usually genes for antibiotic resistance. For example, Tn3 carries a gene for β-lactamase, which confers resistance to ampicillin. **(Procaryotic transposons)**

16. Transposons often carry the genes for antibiotic resistance. Thus, the genes for antibiotic resistance can be spread easily between bacterial DNA molecules. **(Procaryotic transposons)**

17. The enzyme transposase. **(Procaryotic transposons)**

18. In replicative transposition, the transposon leaves a copy of itself at the original site after transposition. Thus, the number of copies of the transposon is increased by one. In conservative transposition, the transposon leaves the original site and moves to a new site in the genome. Thus, the number of transposons does not change. **(General transposons)**

19. The phage mu prophage acts as a replicative transposon, jumping around the bacterial chromosome. **(Procaryotic transposons)**

20. The Ac and Ds elements in maize discovered by Barbara McClintock in the 1940's. **(Eucaryotic transposons)**

21. The Ds element has a 194 bp deletion in the transposase gene and thus produces nonfunctioning transposase. The Ac (short for "activator") element provides functional transposase. **(Eucaryotic transposons)**

22. A dominant allele at the C gene of corn inhibits purple coloration; any kernel possessing the allele is colorless. Some of the kernels McClintock examined were mosaics of purple and colorless. This results from frameshift mutations that inactivate the dominant allele. These mutations, of course, were the result of transposition. **(Eucaryotic transposons)**

23. F. In *Drosophila*, transposons make up 1% of the genome. **(Eucaryotic transposons)**

24. They are essentially the same, except that the Ds element has a 194 bp deletion in its transposase gene. **(Eucaryotic transposons)**

25. A retrotransposon is a transposon that moves using the process of reverse transcription catalyzed by the enzyme reverse transcriptase. Basically, once the retrotransposon is transcribed, the RNA is reverse transcribed to double-stranded DNA and the DNA inserted into the genome at another site. **(Eucaryotic transposons)**

26. A pseudogene is a DNA sequence that is clearly homologous to a gene (i.e., it has a promoter, exons, introns, etc.), but that has lost the ability to be expressed. A processed pseudogene is a pseudogene that appears to have resulted from reverse transcription followed by insertion of an mRNA transcript of a normal gene. The absence of introns and the presence of poly(A) tails provides evidence that processed pseudogenes were once mRNAs. **(Eucaryotic transposons)**

27. If two transposons pair up and undergo recombination, chromosomal rearrangements can result. If the cross-over occurs between two transposons with the same orientation, a deletion will result. If the cross-over occurs between two transposons in opposite orientation, an inversion will result. These events are examples of ectopic intrachromosomal exchanges. Ectopic interchromosomal exchanges are also possible. There is some evidence that this has occurred in *Drosophila*. **(Eucaryotic transposons)**

28. Hybrid dysgenesis results when P strain males mate with M strain females. The result is that the offspring suffer from severe DNA damage. The P strain males carry a type of transposon known as the P element. Normally, movement of the P elements is repressed. When the M cytotype is combined with the P cytotype in an offspring, somehow repression of P transposition is removed and mutations result. **(Eucaryotic transposons)**

Grade Yourself

Circle the numbers of the questions you missed, then fill in the total incorrect for each topic. If you answered more than three questions incorrectly, you need to focus on that topic. (If a topic has less than three questions and you had at least one wrong, we suggest you study that topic also. Read your textbook, a review book, or ask your teacher for help.)

Subject: Transposable Elements

Topic	Question Numbers	Number Incorrect
General transposons	1, 18	
Procaryotic transposons	2, 3, 4, 5, 6, 7, 8, 9, 10, 11, 12, 13, 14, 15, 16, 17, 19	
Eucaryotic transposons	20, 21, 22, 23, 24, 25, 26, 27, 28	

Developmental Genetics

19

 Brief Yourself

Development

Development in eucaryotes involves two processes: cell division and cell differentiation. Once an egg is fertlilized to form the zygote, it begins to divide rapidly eventually to produce the billions or even trillions of cells that comprise an adult. Of course, cell division is tightly regulated; many genes are devoted to control of the cell cycle. Mutations that disable these regulatory genes can result in cancer, a disastrous disease characterized by runaway cell division. During development, gene expression is regulated in two dimensions: temporally and spatially. An example of temporal gene regulation during development is the production of hemoglobin, the protein that is responsible for transporting oxygen in the blood of vertebrates. Various globin genes are expressed at different times during development, apparently because different developmental stages require slightly different hemoglobins. Cell differentiation results from spatial control of gene expression. An example of spatial regulation would be the differences between a liver cell and a brain cell. Because each type of cell has very different tasks, they express different (but overlapping) sets of genes.

Model Organisms

In recent years, the organisms of choice for developmental geneticists have become the fruit fly *Drosophila melanogaster* and the nematode *Caenorhabditis elegans*. Both have advantages to offer developmental geneticists. *Drosophila* flies are easy to keep in the lab, have short generation times, only have 4 chromosomes, and have been studied by geneticists for almost 90 years. *Caenorhabditis elegans* is ideal for developmental research because of the simplicity of their development. They are composed of only 959 cells, and very importantly, the history of these cells is completely predictable. The fate of any cell is already known in advance.

Developmental Genetics

The objective of developmental genetics is to understand how genes are regulated during development. It is now well understood that even before cells begin to express the genes associated with their final differentiation, they are already committed to follow a certain pathway. This is known as determination. Determination starts with maternally contributed gene products in the egg. The early course of development is now fairly well understood in *Drosophila*. Initially, development in *Drosophila* involves the establishment of the body axes by maternally contributed morphogens. These morphogens trigger expression of certain genes in a concentration-dependent way. These genes express regulatory proteins that trigger the expression of a new set of regulatory genes. Thus, development involves a cascade of regulatory gene expression. In this way, the developing embryo is divided into finer and finer domains of cells, each differing in the genes they express. Eventually, a fully differentiated adult is produced.

Test Yourself

1. During development, only spatial control of genes occurs. T or F?

2. What is another term for spatial control of gene expression?

3. The only thing affecting the development of an embryo is the embryo's own genes. T or F?

4. There are two possibilities for cell differentiation. Genes might be selectively degraded in different cells, such that a differentiated cell would lack the genes not related to its particular role. Or every cell might contain the same sets of genes, but experience differential expression of those genes. Which is the truth?

5. A cell whose developmental fate is decided is _____.

6. What are perhaps the two most important study organisms for developmental genetics?

7. What are the different levels of gene regulation?

8. The following elements are involved in gene expression. Define each.

   a. promoter

   b. transcription factor

   c. enhancer

   d. silencer

9. Contrast determination and differentiation.

10. What are maternal-effect genes?

11. A mother is + – for a maternal-effect gene. She produces a – – zygote after mating with a + – father. Will the offspring show the dominant, wildtype trait or the recessive, mutant trait?

12. Concentration gradients of regulatory proteins in the *Drosophila* egg establish the body axes. What are these axes?

13. The regulatory proteins that establish the axes are known as _____.

14. If a researcher wanted to study genes controlling the earliest events in *Drosophila* development, what mutant phenotypes should the researcher look for?

15. What happens to *Drosophila* embryos with null mutations affecting the A/P genes?

16. The early *Drosophila* embryo is a syncytium. What does this mean?

17. What happens to *Drosophila* embryos with null mutations affecting the D/V genes?

18. What are two techniques from molecular biology that can be used to localize morphogens in an embryo?

19. What are the two morphogens that establish the A/P axis in *Drosophila*?

20. What is the morphogen that establishes the D/V axis in *Drosophila*?

21. How are the morphogen gradients of the BCD and HB-M proteins established?

22. What genes respond to the BCD, HB-M, and DL transcription factors?

23. The proteins expressed by the cardinal genes lead to localized expression of what set of genes in *Drosophila*?

24. What are the roles for the pair-rule and segment-polarity genes?

25. What is the function of homeotic genes?

26. What feature do all homeotic genes share?

27. Contrast the homeotic genes of insects and mammals.

28. What is true of the order of expression of homeotic genes relative to the A/P axis?

29. Why is development described as a cascade of regulatory events?

30. How is sex determination achieved in *Drosophila*?

31. Explain what the numerator and denominator elements are in *Drosophila* sex determination.

32. What is the function of the *Drosophila* dsx gene?

33. What is the importance of alternate splicing to sex determination in *Drosophila*?

34. Describe sex determination in humans.

35. An organism comprised of clones of cells varying by genotype is a genetic mosaic. What are some causes of genetic mosaicism?

36. What is striking about cell division in *Caenorhabditis elegans* leading to the 959-celled adult?

37. Describe cell division in *Caenorhabditis elegans* leading to the first 6 cells. What genes contribute to the asymmetric divisions?

Check Yourself

1. F. During development, gene expression is regulated in two dimensions: spatial and temporal. Temporal regulation is a major theme, especially during early development when sets of genes are sequentially turned on in a regulatory cascade. **(General developmental genetics)**

2. Cell differentiation. **(General developmental genetics)**

3. F. The genes that affect the formation of the axes in earliest development are maternal-effect genes, so named because they are expressed maternally and then the gene products sequestered into the embryo. **(General developmental genetics)**

4. For the most part, every cell in an adult eucaryote contains the same genes—no DNA is lost (there are certain exceptions). Thus, cell differentiation is due to differential gene expression, rather than differential gene loss. **(General developmental genetics)**

5. Determined. **(General developmental genetics)**

6. *Drosophila* fruit flies and *Caenorhabditis elegans* nematode worms. **(General developmental genetics)**

7. There are four possible levels of regulation of gene expression. First, the processes of transcription and translation can be regulated. Or second, the products of mRNA and proteins can be modified, repressed, or destroyed. **(General developmental genetics)**

8. a. promoter—a sequence adjacent to the gene that is the binding site for RNA polymerase

    b. transcription factor—a trans-acting protein that regulates the transcription of genes

    c. enhancer—a cis-acting sequence that can amplify the rate of transcription of a gene

    d. silencer—a cis-acting sequence that can lower the rate of transcription of a gene.
        (General developmental genetics)

9. Determination is the process whereby a cell's fate is decided; differentiation is the realization of that fate. **(General developmental genetics)**

10. Maternal-effect genes are genes that are maternally expressed and then sequestered into an egg, where they exert control over early development. **(General developmental genetics)**

11. The offspring will show the dominant trait because for maternal-effect genes, only the genotype of the mother matters. **(General developmental genetics)**

12. The anterior-posterior (A/P) and dorsal-ventral (D/V) axes. **(Drosophila developmental genetics)**

13. Morphogens. Morphogens are regulatory (transcription factors) proteins that act in a concentration-dependent way. **(Drosophila developmental genetics)**

14. Infertility. Null mutations to maternal-effect genes will result in infertility. **(Drosophila developmental genetics)**

15. Formation of the anterior structures is prevented. (**Drosophila developmental genetics**)

16. A syncytium is a cell with many nuclei, but no cell membranes separating the nuclei. (**Drosophila developmental genetics**)

17. Formation of the ventral structures is prevented. (**Drosophila developmental genetics**)

18. RNA in situ hybridization and immunohistochemistry can be used to localize the products of morphogenic genes. (**Drosophila developmental genetics**)

19. The BCD and HB-M proteins, products of the bcd and hb-m maternal-effect genes. (**Drosophila developmental genetics**)

20. The DL protein, product of the dl maternal-effect gene. (**Drosophila developmental genetics**)

21. The BCD protein becomes localized anteriorly because its mRNAs are concentrated anteriorly. This results from the mRNAs binding to the anterior (–) ends of polarized microtubules. The HB-M protein gradient is produced by post-transcriptional regulation. The NOS protein (product of the nos gene) is distributed in the opposite gradient, with its greatest concentration at the posterior end. This is achieved by preferential binding of the NOS mRNAs to the posterior (+) ends of polarized microtubules. The NOS protein represses translation of the HB-M mRNA, resulting in a greater concentration of the HB-M protein anteriorly. (**Drosophila developmental genetics**)

22. The cardinal genes. (**Drosophila developmental genetics**)

23. Secondary regulatory genes. (**Drosophila developmental genetics**)

24. The pair-rule and segment-polarity genes are secondary regulatory genes for the A/P axis in *Drosophila*. The pair-rule genes determine the correct number of segments, while the segment-polarity genes further refine the work started by the pair-rule genes. (**Drosophila developmental genetics**)

25. Homeotic genes in general determine segmental identity in animals. In *Drosophila*, they are secondary regulatory genes of the A/P axis. (**Drosophila developmental genetics**)

26. Homeotic genes code for transcription factors containing a 60 amino acid (coded by the 180 bp homeobox) domain called the homeodomain that is the site of DNA binding. (**Homeotic genes**)

27. In insects, the homeotic genes are referred to as the HOM-C genes and there is only one cluster. In mammals, the homeotic genes are referred to as the hox genes and there are four clusters, the result of duplications. (**Homeotic genes**)

28. The homeotic genes are ordered from anterior to posterior on chromosomes according to the position of their expression on the A/P axis. The mechanism behind this interesting fact is still not understood. (**Homeotic genes**)

29. Because regulatory genes turned on early in development lead to the expression of additional sets of regulatory genes, etc. (**General developmental genetics**)

30. By the ratio of the number of X chromosomes to the autosomes. If a fly is XX, the ratio is 1 (the number of autosomes is 2) and the fly becomes a female. If the fly is XY, the ratio is 0.5 and the fly becomes a male. (**Drosophila sex determination**)

31. The numerator and denominator elements in *Drosophila* are the elements that measure the ratio of X chromosomes to autosomes. Numerator elements are coded by genes on the X chromosome; denominator elements are coded by genes on the autosomes. If numerator elements exceed the denominator elements, the Sxl gene is activated and the fly develops into a female. (**Drosophila sex determination**)

32. The dsx gene codes for the DSX protein, which suppresses the genes required for male development. Thus, expression of the dsx gene causes the embryo to develop into a female. (**Drosophila sex determination**)

33. Alternate splicing is a mechanism of gene regulation at the mRNA level, whereby certain exons can be deleted in different ways to produce different proteins. In *Drosophila*, the SXL protein regulates the splicing of itself and the tra gene product. This leads to functional TRA protein which in turn, along with the TRA2 protein, regulates the splicing of the dsx gene. (**Drosophila sex determination**)

34. Sex determination in humans is much simpler than in *Drosophila*. Rather then being determined by the number of X chromosomes, sex in humans is determined by the presence or absence of the Y chromosome. A gene on the Y chromosome codes for the testis-determining factor, a protein that triggers the expression of other genes leading to the development of a male. (**Human sex determination**)

35. Several causes of genetic mosaicism are the following:

    a. Somatic point mutations

    b. Inactivation of the Barr bodies

    c. Somatic recombination

    d. Transposition

    e. Mitotic assortment of organellar genetic variation. (**Genetic mosaicism**)

36. That cell division is deterministic, meaning that it is completely predictable. Given any cell in the development of *Caenorhabditis elegans*, the cells it will give rise to can be predicted. (**Caenorhabditis elegans developmental genetics**)

37. Each of the initial divisions is asymmetric; each division leads to unequally sized cells, the smaller containing most of the P granules. The P granules lead to differentiation of some of the cells as germ cells. The par genes are responsible for the asymmetric division. (**Caenorhabditis elegans developmental genetics**)

Grade Yourself

Circle the numbers of the questions you missed, then fill in the total incorrect for each topic. If you answered more than three questions incorrectly, you need to focus on that topic. (If a topic has less than three questions and you had at least one wrong, we suggest you study that topic also. Read your textbook, a review book, or ask your teacher for help.)

Subject: Developmental Genetics

Topic	Question Numbers	Number Incorrect
General developmental genetics	1, 2, 3, 4, 5, 6, 7, 8, 9, 10, 11, 29	
Drosophila developmental genetics	12, 13, 14, 15, 16, 17, 18, 19, 20, 21, 22, 23, 24, 25	
Homeotic genes	26, 27, 28	
Drosophila sex determination	30, 31, 32, 33	
Human sex determination	34	
Genetic mosaicism	35	
Caenorhabditis elegans developmental genetics	36, 37	

Extranuclear Inheritance

20

Brief Yourself

Eucaryotic organelles arose through endosymbiosis more than a billion years ago. The origin of the mitochondrion came first and involved the invasion of a large anaerobic cell by a smaller aerobic cell to form a stable endosymbiosis. The aerobic cell eventually evolved into the mitochondrion, which is the site of oxidative phosphorylation in eucaryotes. Later, a photosynthetic bacterium invaded a lineage of eucaryotes and formed a stable endosymbiosis. Of course, these eucaryotes evolved into the photosynthetic eucaryotes and the endosymbiosed cells developed into chloroplasts. Because mitochondria and chloroplasts represent ancient endosymbioses of formerly free-living cells, they contain their own chromosomes. This complicates the study of heredity, because one must now consider mitochondria and chloroplasts in addition to the nucleus as the possible sources of phenotypes. There are two major differences in the heredity of organellar genes compared to nuclear genes: first, organellar DNA is (usually) haploid, and second, organellar genes are passed on only by the mother of an offspring. Therefore organellar genes show maternal inheritance; even in species with no separate sexes, the organelles are passed on to the offspring by only one parent.

Test Yourself

1. In four-o'clock plants, there are two varieties: variegated and green. When male variegated plants are crossed with female green plants, a different result is obtained than when female variegated plants are crossed with male green plants. What is this due to?

2. Why are some four-o'clock plants variegated?

3. Are all traits that are maternally inherited also organellar?

4. Can gene mapping of organellar genes be performed by analysis of the products of recombination?

5. In the plant *Pelargonium zonale*, if a green ovule is crossed with a white pollen grain, a mixture of offspring is produced consisting of green, variegated, and white plants. Explain this result.

6. What is the range of variation in the size of the mitochondrial genome?

7. All mitochondrial DNAs (mtDNAs) are circular. T or F?

8. Sometimes organellar genes can exist on several chromosomes. T or F?

9. What are some aspects of mtDNA that distinguish it from nuclear DNA (nDNA)?

10. Human beings have 37 mtDNA genes. What is the breakdown of these genes by type?

11. What is a URF?

12. How do the two strands of an mtDNA molecule differ?

13. Describe the transcription of the human mtDNA molecule.

14. How does translation of mtDNA mRNAs differ from those of nuclear genes?

15. Some mtDNA genes carry introns. T or F?

16. How is plant mtDNA gene expression different from that of animal mtDNA?

17. Explain the phenomenon of trans splicing.

18. Are mitochondria autonomous from nuclear genes?

19. What are some genetic diseases in humans due to mtDNA mutations?

20. What characterizes people with Pearson marrow-pancreas syndrome?

21. How large are chloroplast DNA (cpDNA) molecules?

22. What type of genes are carried by cpDNA?

23. What evidence links mitochondria and chloroplasts to their bacterial ancestors?

24. What type of inheritance is suggested by the following pedigree for a rare genetic disease?

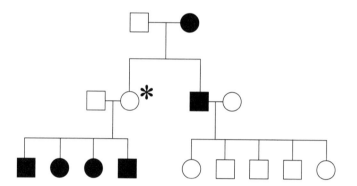

✓ Check Yourself

1. This is maternal inheritance. Usually maternal inheritance is mediated by organellar DNA. **(Extranuclear inheritance)**

2. Because these plants are heteroplasmic, having both mutant chloroplasts incapable of making chlorophyll and normal chloroplasts that make chlorophyll. Mitotic segregation of these variant chloroplasts will result in sectors of cells that lack chlorophyll and others that have chlorophyll. **(Extranuclear inheritance)**

3. No. For example, in the snail genus *Limnaea*, the direction of shell coiling is maternally inherited but determined by nuclear genes. **(Extranuclear inheritance)**

4. Yes. If cells are heteroplasmic, recombination can take place between homologous sites to produce recombinant organellar DNAs. **(Extranuclear inheritance)**

5. The male contributes chloroplasts as does the female. Thus, although organelles are generally maternally inherited, there are exceptions. **(Extranuclear inheritance)**

6. From 16 kb in mammals to 2500 kb in flowering plants. **(mtDNA)**

7. F. Some organisms such as the alga *Chlamydomonas* have linear mtDNAs. (**mtDNA**)

8. T. Occasionally, intramolecular recombination will break a single organellar DNA up into several molecules. (**mtDNA**)

9. mtDNA is (usually) circular, free of histones, (usually) free of introns, and possesses very little intergenic DNA. (**mtDNA**)

10. The 37 mtDNA genes in humans are divided among three classes:

    a. 22 tRNA genes

    b. 2 rRNA genes

    c. 13 protein-coding genes. (**mtDNA**)

11. URF stands for "unassigned reading frame", which is a gene whose product is unknown. (**mtDNA**)

12. One strand—the H (for "heavy")—is heavier than the other, called the L (for "light") strand. This is due to a preponderance of heavy nucleotides on the H strand. (**mtDNA**)

13. The two strands are transcribed as a unit from a promoter upstream of the phe tRNA. Once transcription is initiated, it proceeds in opposite directions on both strands until termination. The polygenic RNAs are then cleaved, polyadenylated, and translated. (**mtDNA**)

14. Translation of mtDNA genes is basically the same as it for nDNA genes, with two exceptions: first, mitochondrial ribosomes are smaller than nuclear ribosomes, and second, some mtDNA codons code for different amino acids than their nuclear counterparts. (**mtDNA**)

15. T. For example, cytochrome oxidase subunit I of yeast has 9 introns. (**mtDNA**)

16. Plant mtDNA gene expression is different from that of animal mtDNA in several ways, including:

    a. plant genes are transcribed into many transcription units,

    b. mRNA editing occurs in plants, and

    c. trans splicing occurs. (**mtDNA**)

17. Trans splicing is the process whereby several exons are transcribed separately and then spliced together prior to translation. (**mtDNA**)

18. No. Many of the functions carried out by mitochondria, such as transcription and translation of mtDNA genes and oxidative phosphorylation, depend on nuclear genes. In some cases, a protein has both mitochondrial and nuclear subunits. (**mtDNA**)

19. Several human genetic diseases due to mutations to mtDNA genes are: LHON (Leber's hereditary optic neuropathy), and NARP (neurogenic muscle weakness, ataxia, and retinitis pigmentosum). All of these are due to mutations to the proteins involved in oxidative phosphorylation. (**mtDNA**)

20. People with Pearson marrow-pancreas syndrome lose bone marrow cells during childhood, which is frequently fatal. The disease is due to a deletion mutation to the mtDNA. Actually, people with the disease are heteroplasmic, possessing some normal mitochondria. The reason for this is that people homoplasmic for the mutant mtDNA fail to develop. (**mtDNA**)

21. They range from about 100 to 200 kb, although some plants have much larger cpDNAs. (**cpDNA**)

22. The genes carried by cpDNA fall into several categories:

     a. rRNA genes

     b. tRNA genes

     c. Ribosomal protein genes

     d. RNA polymerase genes

     e. Photosystem genes

     f. Carbon fixation genes. (**cpDNA**)

23. Several lines of evidence link mitochondria and chloroplasts to their origins as endosymbiotic bacteria:

     a. Circular DNAs in mitochondria and chloroplasts, just like in bacteria,

     b. Mitochondria and chloroplasts have their own ribsomes, which are similar to those of bacteria,

     c. DNA organization: no histones, few introns, little intergenic DNA, and

     d. Similar rRNA genes. (**Origins of mitochondria and chloroplasts**)

24. This pedigree is consistent with extranuclear inheritance. It might be consistent with a nuclear dominant allele, except that then one would have to explain why the female marked with an asterisk lacked the trait. Under the extranuclear hypothesis, this female's lack of the trait can be explained by assuming that she was heteroplasmic for the mutant and normal organelles. Under this hypothesis, we would assume that the original female was heteroplasmic, but that she passed on a gamete to her daughter that contained predominantly the normal organelle. However, by mitotic assortment of the two organelles, gametes were generated by the daughter containing mainly the mutant organelle. (**Extranuclear inheritance**)

Grade Yourself

Circle the numbers of the questions you missed, then fill in the total incorrect for each topic. If you answered more than three questions incorrectly, you need to focus on that topic. (If a topic has less than three questions and you had at least one wrong, we suggest you study that topic also. Read your textbook, a review book, or ask your teacher for help.)

Subject: Extranuclear Inheritance

Topic	Question Numbers	Number Incorrect
Extranuclear inheritance	1, 2, 3, 4, 5, 24	
mtDNA	6, 7, 8, 9, 10, 11, 12, 13, 14, 15, 16, 17, 18, 19, 20	
cpDNA	21, 22	
Origins of mitochondria and chloroplasts	23	

Population Genetics

21

Brief Yourself

Population Genetics

Population genetics is the subdiscipline of genetics that deals with the dynamics of allele frequencies in populations. Because evolution is defined as a change in allele frequency, population genetics is critical to the study of evolutionary biology. For this chapter, it is necessary to introduce some symbolism. Assume that a population carries two alleles *A* and *a* for a gene. The genotypic frequency f_{AA} is the proportion of individuals who have the *AA* genotype, f_{Aa} is the proportion of individuals who have the *Aa* genotype, etc.; the allele frequency f_A is the proportion of all alleles that are *A*. Allele frequencies are calculated from the genotypic frequencies:

$$f_A = f_{AA} + \frac{1}{2} f_{Aa} \text{, and}$$

$$f_a = f_{aa} + \frac{1}{2} f_{Aa} .$$

The meaning of allele frequencies is simple. If there are N individuals in a diploid population, they contain 2N copies for a particular gene. The allele frequency f_A is the proportion of these 2N copies corresponding to *A*.

Hardy-Weinberg Equilibrium

Early in this century, G. H. Hardy and Wilhelm Weinberg formulated a basic principle relating expected genotypic frequencies to allele frequencies under a set of assumptions. This principle, known as Hardy-Weinberg equilibrium, can be summarized by the following equations for a two-allele locus:

$$f_{AA} = p^2; f_{Aa} = 2pq; f_{aa} = q^2,$$

where $p = f_A$, and $q = f_a$. Thus, the expected genotypic frequencies can be predicted from the allele frequencies. When one or more of the underlying assumptions are violated, the equations above will not accurately predict the genotypic frequencies. Evolutionary forces, which represent violations of Hardy-Weinberg equilibrium, are forces that can change allele frequency through time. Thus, Hardy-Weinberg equilibrium provides a null model of the maintenance of genetic variation when there is no evolution.

Test Yourself

1. What are the assumptions behind Hardy-Weinberg (H-W) equilibrium?

2. For a population, there are 1000 *AA*'s, 1200 *Aa*'s, and 900 *aa*'s. Calculate the genotypic and allele frequencies.

3. As molecular techniques for assessing genetic variation became available, what was learned about the prevalence of genetic variation?

4. If p = .60, what will be the frequency of *Aa* and *aa* assuming H-W equilibrium?

5. If f_{AA} = .60, what is the frequency of the other two genotypes assuming H-W equilibrium?

6. Given genotypic frequencies, it is simple to calculate the allele frequencies. Is the reverse true?

7. For a population, there are 3 *AA*'s, 100 *Aa*'s, and 4 *aa*'s. Is this population in H-W equilibrium?

8. Explain your conclusion from question 7.

9. For a triallelic system, give an equation predicting the frequencies of the genotypes under H-W equilibrium.

10. For the ABO blood group, a population has a frequency of .2 for the I^A allele and .3 for the I^B allele. Using the equation from question 9, predict the frequency of each blood phenotype.

11. Given the following data, calculate the allele frequencies for the I^A, I^B, and I^O alleles: 200 type A's, 100 type B's, 20 type AB's, and 200 type O's.

12. Assume that for the population from question 11, f_{AA} = .30; f_{AO} = .08; f_{AB} = .04; f_{BB} = .19; f_{BO} = .0; and f_{OO} = .38. Is the population in H-W equilibrium?

13. Explain natural selection.

14. What are the two components of natural selection?

15. For the following data, calculate the mean fitness:

    f_{AA} = .30, W_{AA} = .9

    f_{Aa} = .30, W_{Aa} = .9

    f_{aa} = .40, W_{aa} = .2.

16. From question 15, what will be the new frequency of *A* after one unit of time?

17. The general equation for change in allele frequency under natural selection is:

    $$\Delta p = \frac{pq(W_A - W_a)}{\overline{W}}$$

    What does this equation predict about the magnitude of allele frequency change?

18. In certain areas of Africa, individuals who are heterozygous for sickle-cell anemia actually enjoy an advantage in survival over those who are homozygous for the normal allele. Why?

19. What type of selection is represented by the example in question 18? What will happen to the frequency of the sickle-cell allele?

20. Describe the theory of multiple adaptive peaks.

21. What maximizes the rate of allele frequency change under genetic drift?

22. What is the founder effect?

23. What is a neutral mutation?

24. What is the net effect of migration?

25. What are two ways that the random mating assumption of H-W equilibrium can be violated?

Check Yourself

1. The assumptions are:

    a. random mating

    b. infinite population size

    c. no selection

    d. no mutation

    e. no migration (**H-W equilibrium**)

2. The genotypic frequencies are: f_{AA} = 1000/3100 = .32; f_{Aa} = .39; f_{aa} = .29. The allele frequencies are: f_A = .32 + 1/2(.39) = .52; f_a = .48. (**General population genetics**)

3. It was learned that there is extensive genetic variation in natural populations, more than was expected. (**General population genetics**)

4. If p = .60, then q = .40. The genotypic frequencies are then easily calculated: f_{AA} = $.60^2$ = .36; f_{Aa} = 2(.6)(.4) = .48; f_{aa} = $.40^2$ = .16. (**H-W equilibrium**)

5. One must first find the allele frequencies by a reverse of H-W equilibrium: $f_A = \sqrt{.60}$ = .77, f_a = 1 − .77 = .23. Then these frequencies are used to calculate the other genotypic proportions: f_{Aa} = 2(.77)(.23) = .35; f_{aa} = $.23^2$ = .05. (**H-W equilibrium**)

6. Only if H-W equilibrium holds. (**H-W equilibrium**)

7. To answer this question, one first needs to calculate the allele frequencies, and then the expected genotypic frequencies are calculated from the allele frequencies. If the observed and expected genotypic frequencies are close, then one can conclude that the population is in H-W equilibrium. In this case, the expected genotypic frequencies are: f_{AA} = $(3/107 + 1/2(100/107))^2$ = $(.50)^2$ = .25; f_{Aa} = .50; f_{aa} = .25. Clearly, these frequencies are far from the observed frequencies and hence we conclude that the population is not in H-W equilibrium. (**H-W equilibrium**)

8. The reason that the population in question 7 deviates from H-W equilibrium is that there is a preponderance of heterozygotes, and too few homozygotes. (**H-W equilibrium**)

9. For a two-allele system, H-W equilibrium equations are based on the expansion of the binomial $(p + q)^2$. For a three-allele system, we can derive an equation by expansion of the trinomial: $(p + q + r)^2 = p^2 + 2pq + q^2 + 2pr + 2qr + r^2$. (**H-W equilibrium**)

10. Using the equation from question 9:

    frequency type A = $f_{AA} + f_{AO}$ = p^2 + 2pq = .04 + .20 = .24
    frequency type B = $f_{BB} + f_{BO}$ = r^2 + 2qr = .09 + .30 = .39
    frequency type AB = f_{AB} = 2pr = .12
    frequency type O = f_{OO} = q^2 = .25. (**H-W equilibrium**)

11. First, since $f_{OO} = q^2 = 200/520 = .38$, $q = .62$. The frequency of type A's plus type O's is:

$$p^2 + 2pq + q^2 = (p + q)^2.$$

Plugging in numbers,

.77 (the combined frequency of the A's and O's) $= (p + .62)^2$.

Taking the square root of both sides gives us:

.89 = p + .62
p = .26, and hence, r = .12. (**H-W equilibrium**)

12. We can take a shortcut and answer this question based only on f_{OO}. The expected $f_{OO} = q^2$ (the observed frequency of the O allele squared) $= .42^2 = .17$, which is much less than the observed frequency of .38. Thus, this population is not in H-W equilibrium. (**H-W equilibrium**)

13. Natural selection is an evolutionary force stemming from the simple fact that some genetic variants predispose their bearers to increased survival and/or reproduction compared to other variants. When this is true, the allele will increase in frequency in the population. Eventually, the allele can go to fixation (frequency = 100%). (**Natural selection**)

14. Increased survival and/or increased reproduction. Both components will lead to an increase in allele frequency. (**Natural selection**)

15. The mean fitness is calculated by the following equation:

$$\overline{W} = f_{AA}(W_{AA}) + f_{Aa}(W_{Aa}) + f_{aa}(W_{aa}) =$$
$$\overline{W} = .20(.9) + .50(.9) + .3(.2) = .69.$$

(**Natural selection**)

16. The new frequency is calculated according to the following equation:

$$p' = \frac{p(pW_{AA} + qW_{Aa})}{\overline{W}},$$
$$p' = .59.$$

(**Natural selection**)

17. It means that the rate of change of allele frequency will depend on the difference between the fitnesses of the two alleles as well as the factor pq, which is proportional to the frequency of the heterozygotes. In other words, the more balanced are the allele frequencies, the faster *A* will increase in frequency. (**Natural selection**)

18. Because heterozygotes are protected from malaria. (**Natural selection**)

19. Balancing selection. Under balancing selection, a stable equilibrium will develop between the two alleles; neither will go to fixation. (**Natural selection**)

20. Often many genotypes produce the same phenotype, especially for traits coded by multiple genes. If a trait of this nature is favored by natural selection, that trait will define multiple adaptive peaks, each defined by a different genotype. Depending on their starting allele frequencies, different populations will ascend different adaptive peaks (go to fixation for alleles yielding the favored phenotype), and hence develop different genotypes. (**Natural selection**)

21. Small population size. **(Genetic drift)**

22. The founder effect is strong genetic drift that occurs whenever a very small number of individuals founds a new, isolated population. **(Genetic drift)**

23. A neutral mutation is one that confers upon its bearer the same fitness as the wildtype allele. Genetic drift can cause a neutral mutation to go to fixation. **(Genetic drift)**

24. Migration has a homogenizing effect on populations. By constantly exchanging new alleles among different populations, migration counteracts the divergence among populations due to evolutionary forces. **(Migration)**

25. Random mating is violated when matings are biased by the degree of relationship. Inbreeding occurs between related individuals; outbreeding occurs between unrelated individuals. Another way that random mating is violated is assortative mating, mating that is biased by similarity. Positive assortative mating occurs when individuals prefer similar mates; negative assortative mating is the opposite. **(H-W equilibrium)**

Grade Yourself

Circle the numbers of the questions you missed, then fill in the total incorrect for each topic. If you answered more than three questions incorrectly, you need to focus on that topic. (If a topic has less than three questions and you had at least one wrong, we suggest you study that topic also. Read your textbook, a review book, or ask your teacher for help.)

Subject: Population Genetics

Topic	Question Numbers	Number Incorrect
H-W equilibrium	1, 4, 5, 6, 7, 8, 9, 10, 11, 12, 25	
General population genetics	2, 3	
Natural selection	13, 14, 15, 16, 17, 18, 19, 20	
Genetic drift	21, 22, 23	
Migration	24	